A Journey into Reciprocal Space

A crystallographer's perspective

A Journey into Reciprocal Space

A crystallographer's perspective

A M Glazer

Physics Department, University of Oxford and Physics Department, University of Warwick and Jesus College Oxford

Morgan & Claypool Publishers

Copyright © 2017 Morgan & Claypool Publishers

All rights reserved. No part of this publication may be reproduced, stored in a retrieval system or transmitted in any form or by any means, electronic, mechanical, photocopying, recording or otherwise, without the prior permission of the publisher, or as expressly permitted by law or under terms agreed with the appropriate rights organization. Multiple copying is permitted in accordance with the terms of licences issued by the Copyright Licensing Agency, the Copyright Clearance Centre and other reproduction rights organisations.

Rights & Permissions
To obtain permission to re-use copyrighted material from Morgan & Claypool Publishers, please contact info@morganclaypool.com.

ISBN   978-1-6817-4621-0 (ebook)
ISBN   978-1-6817-4620-3 (print)
ISBN   978-1-6817-4623-4 (mobi)

DOI   10.1088/978-1-6817-4621-0

Version: 20171001

IOP Concise Physics
ISSN 2053-2571 (online)
ISSN 2054-7307 (print)

A Morgan & Claypool publication as part of IOP Concise Physics
Published by Morgan & Claypool Publishers, 1210 Fifth Avenue, Suite 250, San Rafael, CA, 94901, USA

IOP Publishing, Temple Circus, Temple Way, Bristol BS1 6HG, UK

Dedicated to all those undergraduates who have had to suffer my tutorials for the last 40 years!

Contents

Preface		x
Acknowledgements		xi
Author biography		xii
1	**Direct space**	**1-1**
1.1	What are crystals?	1-1
1.2	Miller indices	1-2
1.3	Point symmetry	1-4
1.4	Point groups	1-11
1.5	Translational symmetry	1-13
1.6	Crystal structures	1-25
1.7	Space groups	1-33
	References	1-37
2	**The reciprocal lattice**	**2-1**
	Brief history	2-1
2.1	Definition	2-2
2.2	Construction	2-3
2.3	Geometrical calculations	2-6
	References	2-10
3	**Diffraction**	**3-1**
3.1	Laue equations	3-1
3.2	Bragg's Law	3-2
3.3	The Ewald sphere	3-6
3.4	Lost in reciprocal space?	3-6
3.5	Intensity	3-16
3.6	Fourier transformation	3-19
3.7	Convolution theorem	3-20
3.8	Two simple 'Rules'	3-25
3.9	Lattice diffraction	3-26
3.10	Structure factors	3-28
3.11	Form factors	3-29
3.12	Anomalous dispersion	3-31

3.13	Intensity calculations	3-38
3.14	Solution of crystal structures	3-42
3.15	Fourier synthesis	3-44
3.16	The Patterson method	3-48
3.17	Charge flipping	3-50
3.18	The Rietveld method	3-52
3.19	Total scattering analysis	3-54
3.20	Aperiodic crystals	3-55
3.21	Disordered crystals	3-60
	References	3-61

4 Dynamical diffraction — 4-1

4.1	Multiple scattering	4-1
4.2	Renninger effect	4-2
4.3	Two-beam approximation in electron diffraction	4-3
4.4	Pendellösung or thickness fringes	4-8
	References	4-10

5 Waves in a periodic medium — 5-1

5.1	Waves in space	5-1
5.2	Periodic boundary conditions	5-2
5.3	Bloch's theorem	5-4
5.4	Brillouin zones	5-5
5.5	Wigner–Seitz cell	5-7
5.6	Higher-order Brillouin zones	5-10
5.7	Density of states	5-12
	References	5-14

6 Thermal and electronic properties — 6-1

6.1	Specific heat capacity of solids	6-1
6.2	Einstein model	6-2
6.3	Debye model	6-3
6.4	Vibrations of atoms	6-9
6.5	Lattice dynamics	6-21
6.6	Heat conduction	6-23
6.7	Interaction with radiation	6-27

6.8	Free electrons in a metal	6-33
6.9	Nearly free electrons	6-35
6.10	Metal or insulator?	6-38
	References	6-42

Appendix   Wigner–Seitz constructions   **A-1**

Preface

The concept of reciprocal space is over 100 years old, and has been made particular use of by crystallographers in order to understand the patterns of spots when x-rays are diffracted by crystals. However, it has a much more general use, especially in the physics of the solid state. In order to understand what it is, how to construct it and how to make use of it, it is first necessary to start with the so-called real or direct space and then show how reciprocal space is related to it. Real space describes the objects we see around us, especially with regards to crystals, their physical shapes and symmetries and the arrangements of atoms within: the so-called crystal structure. Reciprocal space on the other hand deals with the crystals as seen through their diffraction images. Indeed, crystallographers are accustomed to working backwards from the diffraction images to the crystal structures, which we call crystal structure solution. In solid state physics, one usually works the other way, starting with reciprocal space to explain various solid-state properties, such as thermal and electrical phenomena.

In this book, I start with the crystallographer's point of view of real and reciprocal space and then proceed to develop this in a form suitable for physics applications. Note that while for the crystallographer reciprocal space is a handy means of dealing with diffraction, for the solid-state physicist it is thought of as a way to describe the formation and motion of waves, in which case the physicist thinks of reciprocal space in terms of momentum or wave-vector **k**-space. This is because, for periodic structures, a characteristic of normal crystals, elementary quantum excitations, e.g. phonons and electrons, can be described both as particles and waves. The treatment given here, will be by necessity brief, but I would hope that this will suffice to lead the reader to build upon the concepts described. I have tried to write this book in a suitable form for both undergraduate and graduate students of what today we call 'condensed matter physics'.

Acknowledgements

I was fortunate to learn all about crystallography from two great scientists, Kathleen Lonsdale, my PhD supervisor, and Helen D Megaw at the Cavendish Laboratory, Cambridge, who introduced me to the world of perovskites. Although I had started out as a chemist, I soon appreciated from them the value of condensed matter physics and how it related to crystallography. Thus, when I was appointed to the Clarendon Laboratory, Oxford in 1976 I was in a good position to teach undergraduates and graduates about the solid state, sometimes from a unique point of view. The topics described in this book owe much to hours of discussions, sometimes quite heated, with many of my tutees at Jesus College Oxford. There is nothing like teaching students to make one realize how little one really understands about a subject, and how to relearn something that you thought was already done and dusted. Despite retirement, I continue to learn.

Author biography

A M Glazer

Mike Glazer is Emeritus Professor of Physics at the University of Oxford and Jesus College Oxford, and Visiting Professor at the University of Warwick. From 2014 to 2017 he was also Vice-President of the International Union of Crystallography. His PhD research between 1965 and 1968 was under the supervision of Kathleen Lonsdale at University College London, working on the crystallography of organic mixed crystals. In 1968–1969, he was a Fellow at Harvard University, and then from 1969 to 1976 he was at the Cavendish Laboratory, Cambridge. In 1976, he was appointed Lecturer in Physics at the Clarendon Laboratory Oxford and as an Official Fellow and Tutor at Jesus College Oxford. Mike Glazer's research has mainly been in understanding the relationship between physical properties of crystals and their structures. He is perhaps best known for his classification system for tilted octahedra in perovskites. He is also one of the co-founders of Oxford Cryosystems Ltd, which supplies the world market in low-temperature apparatus for crystallographers.

IOP Concise Physics

A Journey into Reciprocal Space
A crystallographer's perspective
A M Glazer

Chapter 1

Direct space

'Je ne te parlerai que cristaux'
   L Pasteur, in a letter to Charles Chappuis from Strasbourg (July, 1850)

In order to set out on our journey, we shall first have to follow Pasteur and talk about crystals in terms of *Direct* (or *Real*) *Space*. Once this is understood, we can then advance to the next leg of the journey that takes us into *Reciprocal Space* itself, the principal aim of this book. This discussion will centre mainly around crystalline materials from the point of view of their symmetries. There are many books (e.g. [1–5]) and web pages that explain these ideas more fully than I shall be able to do here.

1.1 What are crystals?

Crystals (from the Greek κρυσταλλοσ, meaning 'rock crystal' but also 'ice' from κρυσ, 'icy cold, frost') have been known about for centuries as minerals that have naturally-occurring flat faces bearing some sort of symmetry relationship to each other. Probably the most well-known mineral crystal is that of quartz, especially in its clear form called rock crystal (figure 1.1).

Quartz, chemical formula SiO_2, grows naturally as elongated prisms with large faces arranged around a 3-fold axis of symmetry, together with inclined faces top and bottom. This particular mineral exists in two crystalline habits that are mirror images of each other, a symmetry property known as *chirality* (see [6] for an interesting discussion of the errors made in the literature describing quartz).

Definition. *Crystal habit is the formal name for the shape of a crystal. Some examples of crystal habit are acicular (needle-like), prismatic (elongated prisms) and stellate (star-shaped). There are many others.*

Definition. *Crystal morphology is a term that describes the crystal in terms of its habit, defect nature and polymorphism (where a substance can crystallize in different forms, e.g. diamond and graphite).*

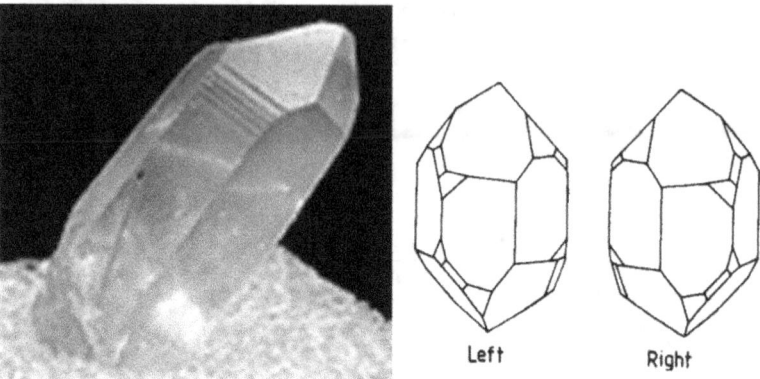

Figure 1.1. On the left an example of a natural crystal of quartz (rock crystal). On the right morphological drawings of left and right-handed habits of quartz crystals.

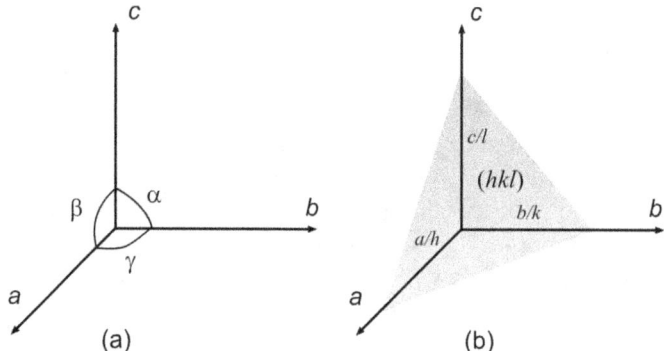

Figure 1.2. Miller indices. (a) Right-hand convention for choice of axes. (b) Definition of a plane.

1.2 Miller indices

In order to be able to characterize the different crystal habits, it is necessary to introduce a system whereby each of the crystal faces can be assigned a label. The most commonly used system was introduced in 1839 by the British mineralogist Wlliam Hallowes Miller. To explain this, we first need to adopt a convention for naming axes and angles in a crystal.

Figure 1.2(a) shows the right-hand screw convention for the choice of axes a, b and c and for interaxial angles α, β and γ. Note that these axes can be any length in general and the angles need not be 90°, depending on the crystal symmetry. In figure 1.2(b), a plane (in grey) intersects the three axes to make intercepts at a/h, b/k and c/l. The plane is then indexed as (hkl). According to the Law of Rational Indices found by René Just Haüy in the 19th century, the indices should be integers, although there are in fact a few rare cases, such as in the mineral calaverite, $AuTe_2$, where some faces could only be indexed on irrational indices.

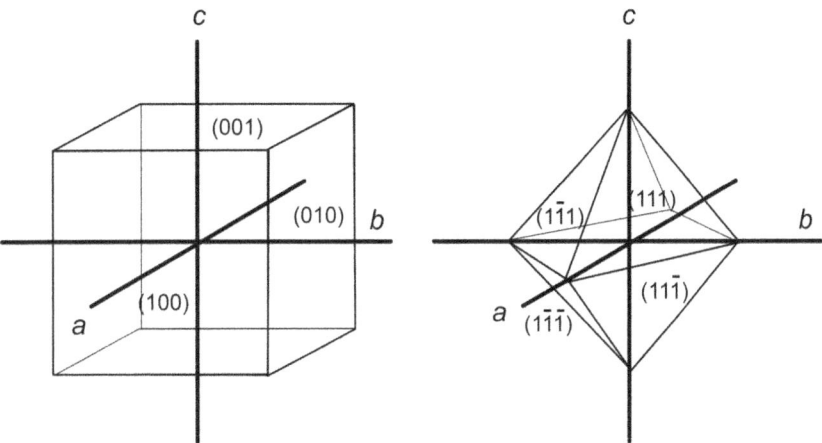

Figure 1.3. Miller indices for the faces of a cube and an octahedron. Only the front faces are marked in this figure.

Notation. *Miller indices are written in parentheses like this (hkl).*

So, a face that is perpendicular to the a axis and is parallel to b and to c has the plane symbol (100), as it cuts off unit intercept along a and infinite intercept along b and c (figure 1.3). Similarly, the faces of an octahedron, once the axes have been assigned, have indices (111), (1$\bar{1}$1), (1$\bar{1}\bar{1}$), (11$\bar{1}$), ... as each face cuts off unit intercept on all three axes.

Notation. *Negative indices are marked with a bar above the number. In my experience, this is said 'bar 1' in the United Kingdom, while Americans usually say '1 bar'! Not sure what the rest of the world says.*

Definition. *The complete set of indexed faces with a particular set of Miller indices related by symmetry is called a form and is denoted {hkl}. Thus, an octahedron has the form {111}.*

Returning to our old friend, quartz (figure 1.4), the habit is complicated and the indices of the faces appear to be unrelated to each other. This arises because of a peculiarity in crystals with so-called hexagonal/trigonal symmetry, i.e., those that contain a single 6-fold or 3-fold axis of symmetry. In this case, the Miller index is best extended to four indices (*hkil*), in other words using four axes rather than three. The c-axis is along the 6- or 3-fold axis, while the other three are perpendicular to c and are 120° from each other. The following rule applies

$$h + k + i = 0. \tag{1.1}$$

So, for example, if we consider the (101) face it can be alternatively written as (10$\bar{1}$1). Then permuting the first three indices we get the {10$\bar{1}$1} form:

$$(10\bar{1}1)\ (\bar{1}011)\ (0\bar{1}11)\ (01\bar{1}1)\ (1\bar{1}01)\ (\bar{1}101) \tag{1.2}$$

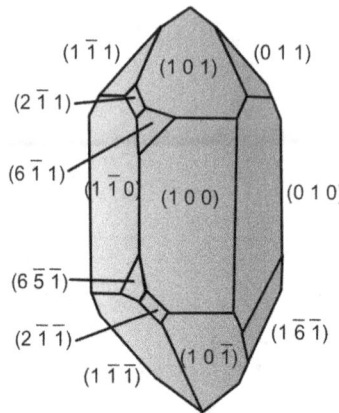

Figure 1.4. Sketch of a quartz crystal showing indices (*hkl*) for the visible faces.

which, when referred back to the 3-axis symbol, becomes

$$(101)\ (\bar{1}01)\ (0\bar{1}1)\ (011)\ (1\bar{1}1)\ (\bar{1}11) \tag{1.3}$$

You can now see why it is in figure 1.4 that the three faces (101), (011), (1$\bar{1}$1) are related about the vertical axis (the three remaining faces of the form are hidden behind the diagram.

In the same way, we see that the two faces marked (6$\bar{1}$1) and (6$\bar{5}$1) (in 4-axis symbols (6$\bar{1}\bar{5}$1) and (6$\bar{5}\bar{1}$1)) are related. as far as the first two indices are concerned, with only the last index being of opposite sign.

1.3 Point symmetry

Tidiness is a virtue, symmetry is often a constituent of beauty…

Winston Churchill

Let us now look at what sort of symmetry operators are needed to describe the symmetry of a crystal.

Definition. *Point symmetry operators are those that describe symmetry operations that act through a point in space.*

In crystallographic symmetry we shall need three types of point symmetry operator: proper rotations, inversions and rotoinversions (or rotoreflections).

Proper rotations

A proper rotation is one that rotates a position vector through an angle about an axis that passes through a defined point. Repeated rotation finally brings the

Table 1.1. Crystallographic rotations.

Symbol	Operation	Element	Matrix
$1(E)$	Identity	–	$\begin{bmatrix} 1 & 0 & 0 \\ 0 & 1 & 0 \\ 0 & 0 & 1 \end{bmatrix}$
$2(C_2)$	2-fold rotation	Axis	$\begin{bmatrix} \bar{1} & 0 & 0 \\ 0 & \bar{1} & 0 \\ 0 & 0 & 1 \end{bmatrix}$
$3(C_3)$	3-fold rotation	Axis	$\begin{bmatrix} 0 & \bar{1} & 0 \\ 1 & \bar{1} & 0 \\ 0 & 0 & 1 \end{bmatrix}$
$4(C_4)$	4-fold rotation	Axis	$\begin{bmatrix} 0 & \bar{1} & 0 \\ 1 & 0 & 0 \\ 0 & 0 & 1 \end{bmatrix}$
$6(C_6)$	6-fold rotation	Axis	$\begin{bmatrix} 1 & \bar{1} & 0 \\ 1 & 0 & 0 \\ 0 & 0 & 1 \end{bmatrix}$

position vector back to its starting position. Table 1.1 shows the rotations (here, about the c-axis) used to describe crystallographic symmetry.

Notice that the only rotations considered are 2, 3, 4 and 6-fold (I have also included operation 1, the trivial identity operation in this table). The symbols are given in two common notations: the International system, described originally by the French mineralogist Charles-Victor Mauguin and the German crystallographer Carl Hermann, and the Schoenflies system (in parentheses). These days, it is the International Notation that is used in crystallography, while subjects such as spectroscopy tend to use the Schoenflies system. The table also gives matrices R that carry out the operation

$$R\mathbf{r} = \mathbf{r}'. \qquad (1.4)$$

Thus, with rotation axes along c the rotation operations have the following mappings of points:

$$\begin{aligned} &2: (x, y, z) \rightarrow (\bar{x}, \bar{y}, z) \\ &3: (x, y, z) \rightarrow (\bar{y}, x - y, z) \\ &4: (x, y, z) \rightarrow (\bar{y}, x, z) \\ &6: (x, y, z) \rightarrow (x - y, x, z). \end{aligned} \qquad (1.5)$$

Note. *The axis of rotation is an example of what is termed a symmetry element, which in the examples given in the table has been chosen to lie along the c-axis of the crystal.*

A useful way to represent the effect of applying symmetry operations is by making use of the stereographic projection. This is a way of representing points on a

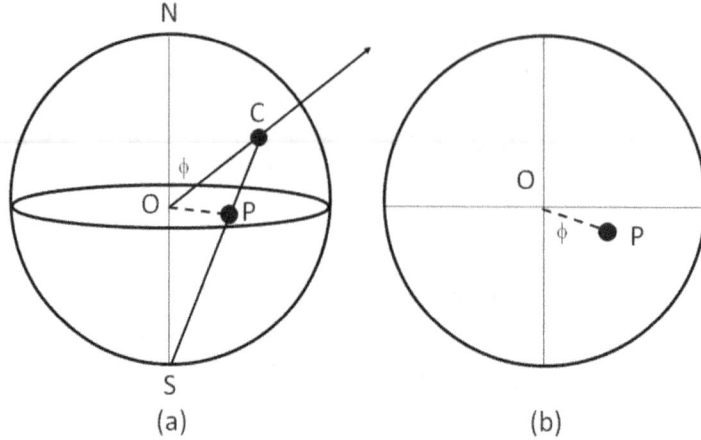

Figure 1.5. The stereographic projection. (a) Construction of a pole (b) resulting projection onto a plane.

sphere by projecting onto a plane, much in the way a map of the world is often represented viewed down onto the North Pole. Imagine (figure 1.5(a)) that there is a point C on the surface of a sphere. Now draw a line SC: this intersects the horizontal plane perpendicular to NS at the point P. N and S can be thought of as the north and south poles, respectively. Figure 1.5(b) shows the projection plane viewed from above where the point P is plotted. The distance OP in this projection is a measure of the angle ϕ measured from the north pole (90°-latitude in cartography) and the point P is called a pole.

Figure 1.6 shows how the effects of symmetry operations are conventionally shown in stereographic projections. The upper four diagrams are for the proper rotations. Consider first the diagram for the rotation operation 2. The small circle with the plus sign next to it represents any object in space (not necessarily a circle!). I shall call this the 'object' and its position in the diagram a general position. The plus sign indicates that it is on the northern hemisphere (you may prefer to think of it as being above the plane of the paper towards you). In each case start with the object near the bottom. The 2-fold operation about the axis perpendicular to the stereographic projection, i.e., along NS, rotates this through 180° to the general position near the top. We see that the resulting object remains in the northern hemisphere as marked by the plus sign. The black symbol at the centre of the stereographic projection, looking somewhat like a rugby football, denotes the 2-fold axis.

Note. *It is conventional to apply rotation operations anticlockwise about the chosen axis.*

Note also, that if the 2-fold rotation is applied again, the object returns to its starting position, i.e.,

$$2^2 = 1 \quad \text{or} \quad C_2^2 = E \tag{1.6}$$

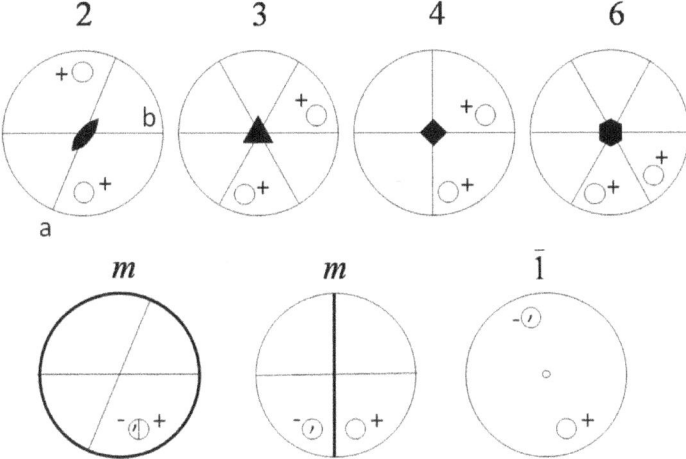

Figure 1.6. Stereographic projections showing rotations and inversion operations.

The next three diagrams show the effect of rotations 3, 4 and 6. By studying these diagrams (or by multiplying out the matrices in table 1.1) we can find interesting relationships such as the following:

$$6^5 = 6^{-1}$$
$$6^4 = 3^2 = 3^{-1} \quad (1.7)$$
$$6^3 = 2$$

Figure 1.7 shows a few examples of rotational symmetry in some everyday objects. It is a good exercise to see how many examples of rotation symmetry you can find during a walk through your neighbourhood.

Reflections and inversions

Symmetry operations that are classed as inversions are given in table 1.2.

The operation of reflection is one that reflects every point in an object as if through a mirror. For a mirror plane perpendicular to the c-axis

$$m: (x, y, z) \rightarrow (x, y, \bar{z}) \quad (1.8)$$

Figure 1.8 also shows some examples of reflections. An important effect of the reflection operation is to convert left-handed objects into right-handed objects, such as in the relationship between the two quartz crystals in figure 1.1.

It is not possible to transform a left-handed object into a right-handed object by using a rotation, and vice versa. Actually, in mathematical terms, this statement is not quite true, provided that we are allowed to change dimensionality: we first transform the description of a three-dimensional chiral object into four dimensions, rotate it and then collapse back into three dimensions! Oh the joys of mathematics!

The bottom line of figure 1.6 shows the stereographic projections for the inversion operations. The first two diagrams are different views of the same type of operation, namely the reflection or mirror operation. In the first stereographic projection, a

Figure 1.7. Some examples of everyday objects showing rotational symmetry. Reprinted from [7] with permission of Elsevier.

Table 1.2. Reflections and inversions.

Symbol	Operation	Element	Matrix
$m(\sigma)$	Reflection	Mirror plane	$\begin{bmatrix} 1 & 0 & 0 \\ 0 & 1 & 0 \\ 0 & 0 & \bar{1} \end{bmatrix}$
$\bar{1}(i)$	Inversion	Inversion centre	$\begin{bmatrix} \bar{1} & 0 & 0 \\ 0 & \bar{1} & 0 \\ 0 & 0 & \bar{1} \end{bmatrix}$

mirror plane is located on the plane perpendicular to the c-axis: this is signified by the thick black line round the perimeter. The effect of this is to reflect the object from the northern to the southern hemisphere (or from above to behind the plane of projection). As this places the reflected object directly beneath the original object in this projection, this is indicated by splitting the object symbol by a vertical line. The minus sign now indicates that the reflected object is below the plane of projection and the comma indicates a change of chirality or hand. The second stereographic projection shows the reflection operation this time with the mirror plane, indicated by the thick line, perpendicular to the b-axis.

The third diagram shows the inversion operation:

$$\bar{1}: (x, y, z) \rightarrow (\bar{x}, \bar{y}, \bar{z}) \tag{1.9}$$

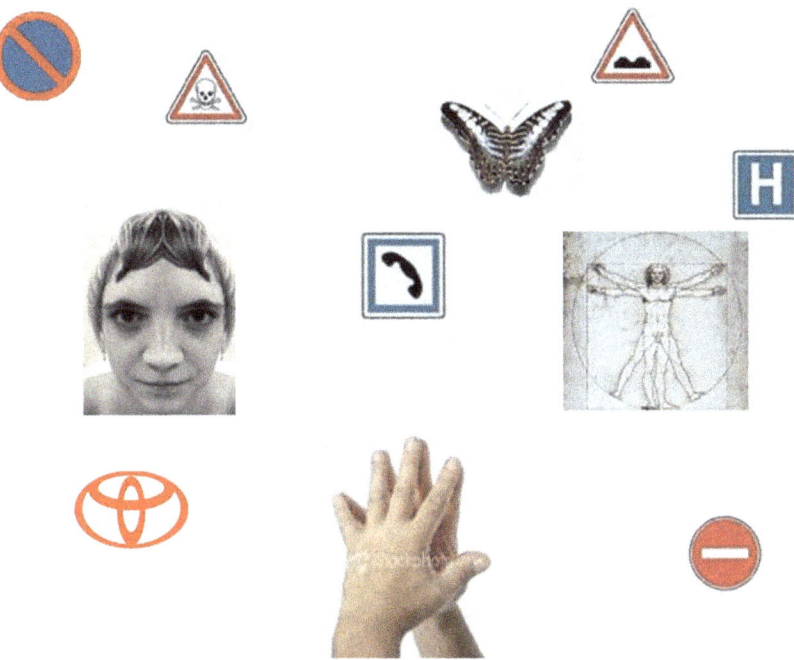

Figure 1.8. Examples of reflection symmetry in everyday objects.

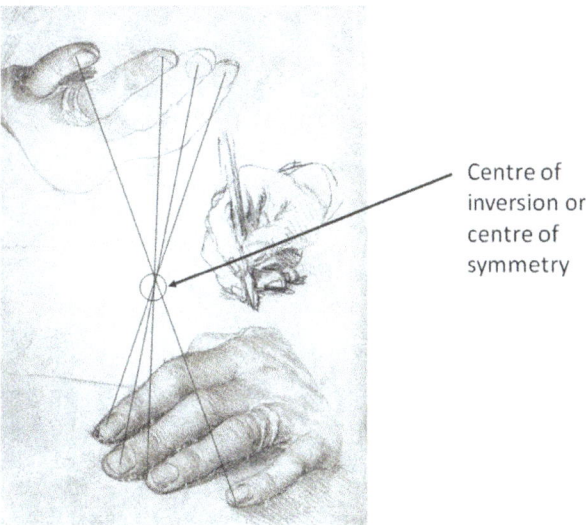

Figure 1.9. Drawing of left and right hands of Erasmus of Rotterdam by Holbein (1523). Lines have been added to indicate the inversion relationship. Reprinted from [7] with permission of Elsevier.

This operation is nicely illustrated by a drawing by Holbein of the hands of Erasmus of Rotterdam (figure 1.9). The relationship between the two hands as drawn shows that each point on one hand is (more or less) related to an equivalent point on the other hand, as indicated by the added lines, and these lines all pass through a point known as the *centre of inversion*, sometimes called the *centre of symmetry*. Any object that contains a centre of symmetry is said to be *centrosymmetric*. Some scientists use the terms *centric* and *acentric*, but this is incorrect, for these terms are used in crystallography to refer to intensity distributions in the diffraction patterns. Note, that just like with reflections, inversion also changes the chirality of an object.

Rotoinversions and rotoreflections

The final types of point symmetry operations that we need to discuss are slightly more complicated than the others. The complication arises because of the different ways in which the International and Schoenflies systems define these operations. These operations are a combination of a rotation and an inversion (International system) or a rotation and a reflection (Schoenflies system).

Table 1.3 lists these operations and figure 1.10 shows the stereographic projections.

Table 1.3. Crystallographic rotoinversions and rotoreflections.

Symbol	Operation	Element	Matrix
$\bar{2}$	Rotoinversion	Point	$\begin{bmatrix} 1 & 0 & 0 \\ 0 & 1 & 0 \\ 0 & 0 & \bar{1} \end{bmatrix}$
S_2	Rotoreflection	Point	$\begin{bmatrix} \bar{1} & 0 & 0 \\ 0 & \bar{1} & 0 \\ 0 & 0 & \bar{1} \end{bmatrix}$
$\bar{3}$	Rotoinversion	Point	$\begin{bmatrix} 0 & 1 & 0 \\ \bar{1} & 1 & 0 \\ 0 & 0 & \bar{1} \end{bmatrix}$
S_3	Rotoreflection	Point	$\begin{bmatrix} 0 & \bar{1} & 0 \\ 1 & \bar{1} & 0 \\ 0 & 0 & \bar{1} \end{bmatrix}$
$\bar{4}$	Rotoinversion	Point	$\begin{bmatrix} 0 & 1 & 0 \\ \bar{1} & 0 & 0 \\ 0 & 0 & \bar{1} \end{bmatrix}$
S_4	Rotoreflection	Point	$\begin{bmatrix} 0 & \bar{1} & 0 \\ 1 & 0 & 0 \\ 0 & 0 & \bar{1} \end{bmatrix}$
$\bar{6}$	Rotoinversion	Point	$\begin{bmatrix} \bar{1} & 1 & 0 \\ \bar{1} & 0 & 0 \\ 0 & 0 & \bar{1} \end{bmatrix}$
S_6	Rotoreflection	Point	$\begin{bmatrix} 1 & \bar{1} & 0 \\ 1 & 0 & 0 \\ 0 & 0 & \bar{1} \end{bmatrix}$

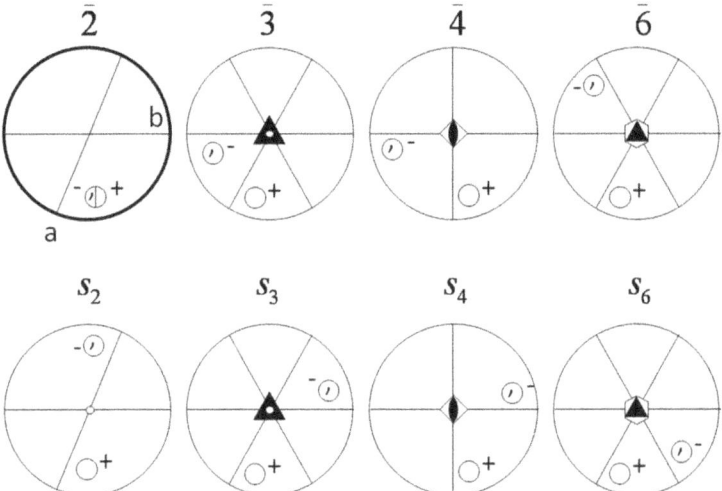

Figure 1.10. Stereographic projections for rotoinversions operations (top) and rotoreflections operations (bottom).

The following mappings then follow for the International system:

$$\begin{aligned}
&\bar{2}: (x, y, z) \rightarrow (x, y, \bar{z}) = m: (x, y, z) \\
&\bar{3}: (x, y, z) \rightarrow (y, \bar{x}+y, \bar{z}) \\
&\bar{4}: (x, y, z) \rightarrow (y, \bar{x}, \bar{z}) \\
&\bar{6}: (x, y, z) \rightarrow (\bar{x}+y, \bar{y}, z)
\end{aligned} \quad (1.10)$$

We see from this several relationships. First of all, the International $\bar{2}$ operation is in fact equivalent to a reflection while the Schoenflies S_2 operation is equivalent to an inversion. More confusingly, perhaps, we find the following:

$$\begin{aligned}
\bar{3} &= S_6^5 = S_6^{-1} \\
\bar{4} &= S_4^3 = S_4^{-1} \\
\bar{6} &= S_3^5 = S_3^{-1} \\
S_3 &= \bar{6}^5 = \bar{6}^{-1} \\
S_6 &= \bar{3}^5 = \bar{3}^{-1}
\end{aligned} \quad (1.11)$$

So, for these types of operations there is a possible confusion between 3-fold and 6-fold roto-operations depending on which system one uses. Beware!

1.4 Point groups

Suppose the symmetry of a crystal can be described by a 2-fold rotation, and at the same time by a reflection perpendicular to the 2-fold axis. Taking the 2-fold axis to lie along the c-axis, and using the matrices in tables 1.1 and 1.2 we find the following relationships.

$$2.2 = 1$$
$$m.m = 1$$
$$2.m = \bar{1} \tag{1.12}$$
$$\bar{1}.\bar{1} = 1$$

This shows that the combination of the 2-fold and perpendicular reflection operation generates a centre of inversion. Therefore, in this case we have four symmetry operations $\{1, 2, m, \bar{1}\}$. We now show that these operations together form a group, as defined in mathematics by the following four criteria.

1. The following multiplication table of operations can be written

	1	2	m	$\bar{1}$
1	1	2	m	$\bar{1}$
2	2	1	$\bar{1}$	m
m	m	$\bar{1}$	1	2
$\bar{1}$	$\bar{1}$	m	2	1

showing that the effect of each operation on another always generates another that is one of the four, and so these operations form a closed set. Because this particular matrix is symmetric, this is known as an Abelian group (not all groups are Abelian though).

2. There is an identify operation 1.
3. There is an inverse operation given by $XX^{-1} = 1$. In this case, each operation is its own inverse.

$$1.1 = 1$$
$$2.2 = 1$$
$$m.m = 1 \tag{1.13}$$
$$\bar{1}.\bar{1} = 1$$

4. Associativity is true. For example

$$2.(m.\bar{1}) = (2.m).\bar{1} \tag{1.14}$$

Figure 1.11 shows a stereographic projection for this group including the symmetry elements. We see from this that this group contains 4 operations, 4 elements and generates 4 general positions and is therefore a group of order 4.

Definition. *The order of a group is given by the number of operations that it contains.*

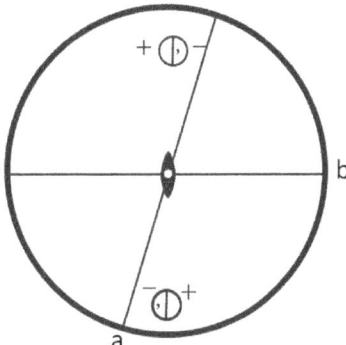

Figure 1.11. Stereographic projection for the point group $2/m(C_{2h})$.

This group is given the symbol $2/m$, the / indicating that the 2-fold axis is perpendicular to the mirror plane, in the International notation. The Schoenflies symbol is C_{2h}: C stands for cyclic and the subscript 2 is for the 2-fold axis and h for a horizontal mirror plane.

Note. *If we use all the possible rotations, inversions and rotoinversion operations relevant to crystal symmetry to form groups it is found that in three dimensions there are 32 distinct point groups, sometimes called the 32 geometric crystal classes. These are listed in table 1.4.*

1.5 Translational symmetry

So far, our discussion on symmetry has been with respect to macroscopic objects such as the crystals we might see around us. We shall now turn our attention to the internal structure of a crystal and how we can describe the arrangements of atoms that make up this structure. The point symmetries will still be relevant at the microscopic scale of atoms and molecules in the crystal. However, before looking at atomic arrangements, we shall first have to introduce a form of symmetry that is characteristic of ideal crystalline materials, *translational symmetry*.

Definition. *Translational symmetry is the symmetry that is exhibited by a collection of equivalent objects repeated regularly throughout space.*

Figure 1.12 shows a picture of a herd of elephants in a regularly repeating pattern that can be described by translational symmetry. Each elephant is at a fixed distance from its neighbour in each direction in space.

Lattices

Suppose now we replace each of the elephants in figure 1.12 by a point, for instance placed at the tip of each tusk. Now remove the elephants (figure 1.13). The result is a

A Journey into Reciprocal Space

Figure 1.12. A herd of elephants illustrating translational symmetry.

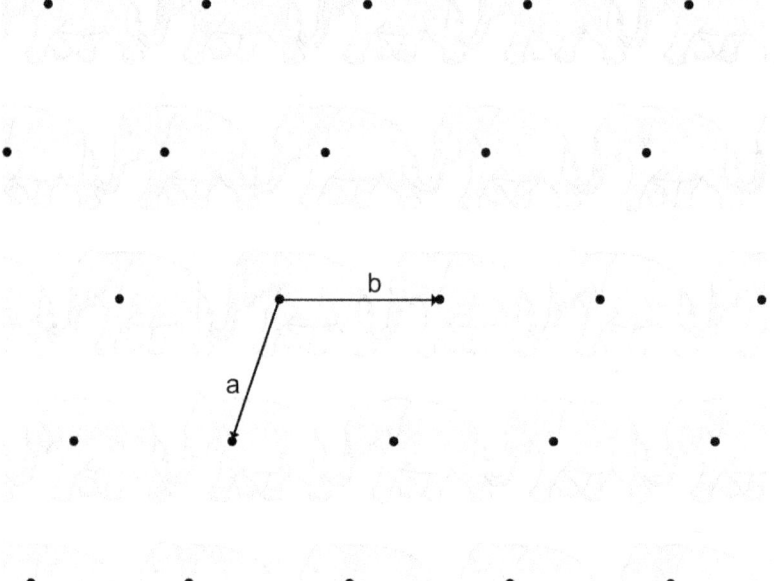

Figure 1.13. Lattice describing translational symmetry of figure 1.12.

repeating pattern of points. Now I have to emphasize something here that I have seen many students (and working scientists for that matter) become confused about. Do not mistake the points in this figure for atoms. These are mathematical, infinitesimal points and the only reason I have drawn them with a finite diameter is so that you can see them. Mathematically we can treat the points as a series of repeating delta functions. This then leads to an important definition.

Definition. *A lattice is a regular repeating array of points (not atoms!).*

Here is my 'Government Health Warning'. Unfortunately, you will see in many textbooks a reference to the word 'lattice' when they actually mean arrays of atoms, not points. So, for instance if we had a repeating pattern of single atoms it would look just like figure 1.13, but in fact this would not be a lattice but a crystal structure (more of this below). In the figure, I have left traces of the original elephants so that you can see where the lattice points have been set. A lattice is a mathematical device that acts as a template to tell us how to position atoms and molecules in space: as such a lattice has no physical existence. A powerful microscope would not show us a lattice but a crystal structure instead. Similarly, when a textbook mentions, for example, the diamond lattice or the copper lattice, they almost certainly should have said the diamond crystal structure and the copper crystal structure. Diamond has pairs of carbon atoms repeating throughout space and copper has single copper atoms repeating throughout space, and so they look quite different. On the other hand, they have the same lattice type, apart from an overall change in dimensions: both have all-face-centred lattices (see below). Please resist using the term 'lattice structures' which you will often see in books: it is an old term and these days it is best avoided as it can create a confusion between lattice and structure. You may think that I am being a bit 'picky', but misunderstandings of this sort can have practical consequences. I recall many years ago, a student who wasted a year's computing resources on an erroneous electronic band structure calculation because he had not understood the difference between lattice and structure.

In figure 1.13 the lattice is shown as a two-dimensional array, but of course in three dimensions there will be repeating lattice points above and below this plane. Now, two arbitrarily chosen axes, a and b, have been marked, with an origin arbitrarily chosen on one of the lattice points (we can assume that there is a c-axis pointing out of the plane of points). Once the axes and origin have been chosen, all lattice points can be reached by the so-called *primitive* translation vector $\mathbf{t}_n$

$$\mathbf{t}_n = n_1\mathbf{a} + n_2\mathbf{b} + n_3\mathbf{c} \tag{1.15}$$

where n_1, n_2 and n_3 are integers. Therefore, it is the primitive translation vector that defines the lattice.

Unit cells

Having defined the lattice, we can make good use of the periodicity it exhibits. Consider figure 1.14, where the same lattice is shown, but this time with two regions outlined in black. These regions are called unit cells.

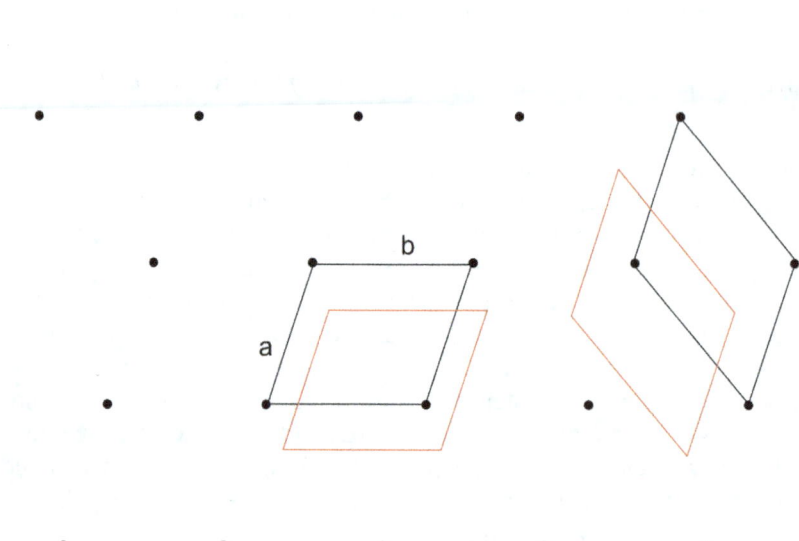

Figure 1.14. A primitive lattice showing primitive unit cells.

Definition. *A unit cell is a region of space which when repeated by primitive translational symmetry fills all space. It can in principle be of any shape, although crystallographers always use a parallepiped shape to describe the unit cell.*

So, the importance of defining a unit cell is that we need only consider the unit cell alone (used in describing crystal structures where we add atoms and molecules into each unit cell) and then allow the primitive translation operator to repeat it. In other words, it is not necessary to write down all the coordinates of the lattice points or atoms in the whole crystal. That after all is the point of symmetry: it enables us to specify just a few things and then allow symmetry operators to generate everything else.

In this lattice two examples of unit cell are drawn, but a moment's thought should convince you that in fact we could define an infinite number of different unit cells. Furthermore, these unit cells both have the same volume (in three dimensions). How do we know this? Simply count the number of lattice points per unit cell and we shall find that it is the same for both unit cells drawn. Now here is a little trick. Often you will see that to count the number of lattice points per unit cell books will often try to partition the lattice points between neighbouring unit cells, as if there were fractions of a lattice point (I don't know how to make a fraction of a point!). So, in the figure we might take a lattice point at each corner of the unit cell and then say that each point is shared between neighbouring unit cells. But there is a much easier way to do the counting. Simply move the origin of the unit cell, as the origin is in fact arbitrary: if we move one unit cell then all the others must move with it and so still fill all space.

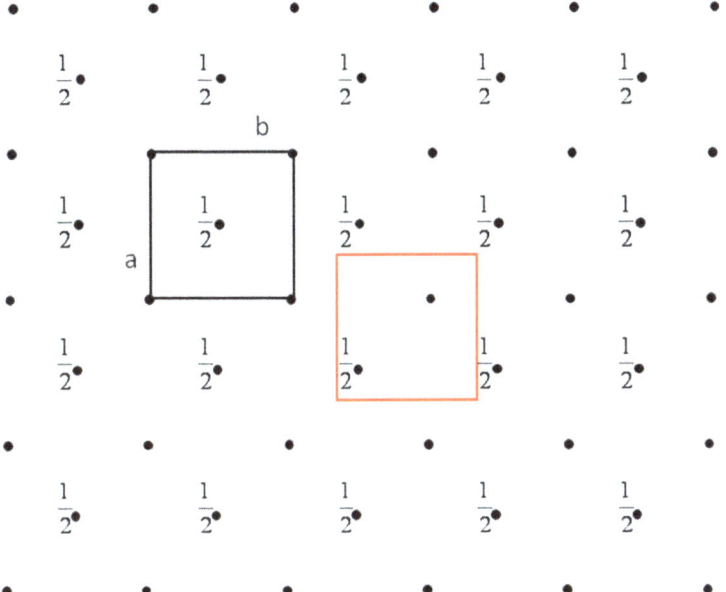

Figure 1.15. Body-centred (*I*) unit cell and lattice. The points marked ½ lie on a plane through $c/2$ above the unmarked points.

Here, the displaced unit cells are in red and then it is easy to see that each contains just one lattice point. These unit cells must therefore have the same volume, like all other possible unit cells that contain just one lattice point.

Definition. *Unit cells that contain one lattice point are called primitive unit cells and the lattice described by them is called a primitive lattice. Primitive lattices are designated by the letter P.*

While in any lattice one can define an infinite variety of primitive unit cells, it is often useful to define so-called *centred* unit cells. In figure 1.15, the unmarked points form a lattice at height 0, while the points marked ½ are those on the plane above at a distance $c/2$. The next plane repeats the plane at height c and so on. Also shown is an example of a unit cell with a lattice point at the origin, at coordinate position (0, 0, 0) and another point in the centre of the unit cell at (½, ½, ½). These coordinates are given as fractions of the unit cell edges a, b and c. Displacing the unit cell (marked in red) immediately shows that this centred unit cell contains two lattice points. Because this unit cell has a lattice point at its centre it is called a *body-centred unit cell* and the lattice is a body-centred lattice. This type of lattice is conventionally given the symbol *I*. Figure 1.16 shows other types of centering together with their conventional symbols and fractional coordinates of the lattice points. The *C*-face centred unit cell can equally be described as *A* or *B*-centering simply by changing the labels for the axes.

A Journey into Reciprocal Space

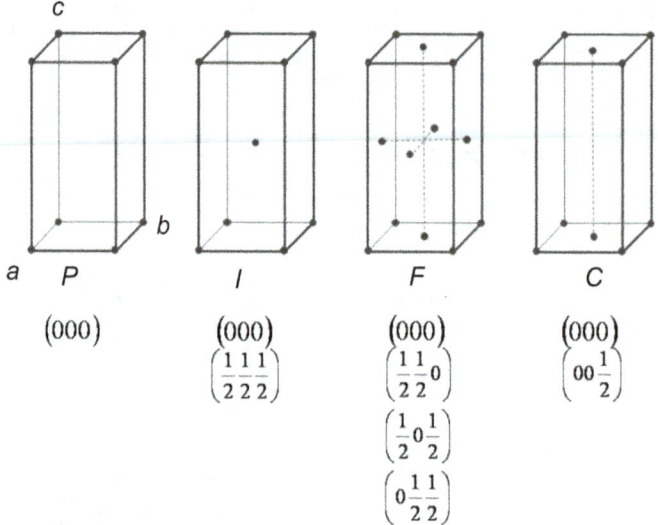

Figure 1.16. Different types of unit cells: *P* primitive, *I* body-centred, *F* all-face centred, *C* one-face-centred. In each case the fractional coordinates of the lattice points are tabulated.

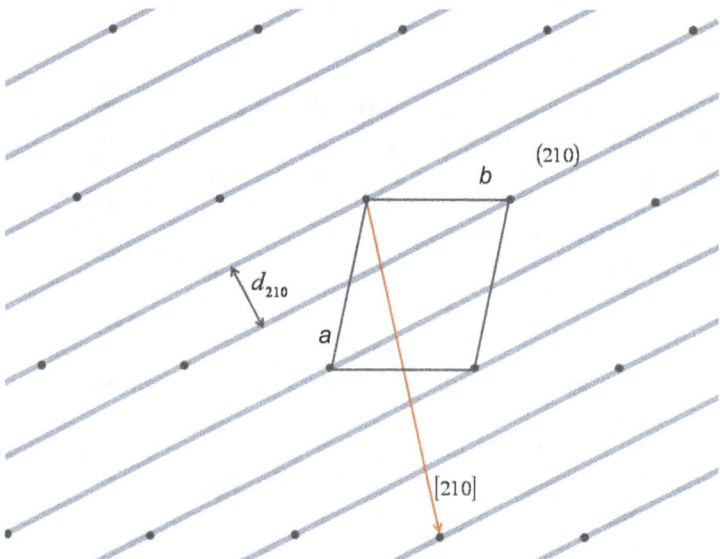

Figure 1.17. Lattice showing (210) planes (blue) and the [210] direction (red).

Planes and directions

We have already seen that Miller indices (*hkl*) are used to denote faces of a crystal. The same types of indices are used to denote planes within a lattice (or for that matter in a crystal structure).

Figure 1.17 shows a section of a lattice on which the (210) plane has been marked in blue: the plane cuts the *a* axis at *a*/2, the *b* axis at unit intercept and is

1-18

parallel to c, out of the plane of the diagram. Now because of translational periodicity, we see that this leads to a whole family of planes separated by the interplanar distance d_{210}. Now, suppose we mark off a vector (drawn in red), from the origin of a unit cell to another lattice point 2 units along a, 1 unit along b and 0 units along c i.e. given by the vector $2\mathbf{a} + \mathbf{b} + 0\mathbf{c}$.

Definition. *A crystal direction is given by $u\mathbf{a} + v\mathbf{b} + w\mathbf{c}$ and is given the symbol [uvw]. The set of directions related by symmetry is denoted by <uvw>.*

Thus, in our example, we denote the direction as [210]. Note that in this case [210] is not perpendicular to the (210) planes: they will be only if the angle between the unit cell axes is 90°. A crystallographic direction is a vector quantity in direct space: we shall see later, when we get to the reciprocal lattice, that it is possible in the case of planes to define another vector, the reciprocal lattice vector, corresponding to a direction *perpendicular* to a plane.

Crystal systems

So far, we have used point symmetry to classify any type of crystal. However, crystals can also be classified into belonging to one of seven so-called crystal systems. They each have a name: cubic, tetragonal, orthorhombic, trigonal, hexagonal, monoclinic and triclinic. These are defined in terms of the minimum symmetry operations and elements that they contain. Table 1.4 lists them together with the effect they have on the unit cell axes and angles. Thus, for example, anything belonging to the cubic system must have four 3-fold axes of symmetry. The effect of this is to permute the a, b and c axes and also ensure that they are orthogonal to each other. Note also that in the monoclinic system there are two common choices for the direction of the 2-fold axis (or perpendicular to the mirror plane). The 1st Setting takes this to be along c, while the more commonly used 2nd Setting by crystallographers, shown here, takes it along b.

Now here is another 'Government Health Warning'. Often in text books you will see that the cubic system is *defined* by $a = b = c$; $\alpha = \beta = \gamma = 90°$. Similarly, the other crystal systems are sometimes defined according to their unit cell geometry. However, this works for defining the crystal systems for lattices, but it should not be used for crystal structures. The restrictions on the lattice parameters in, say, the cubic system are a *consequence* of the four 3-fold axes, and not the other way round. It is perfectly possible in principle to measure the unit cell axes and angles of a crystal and find to within the precision of the measurements relationships like $a = b = c$; $\alpha = \beta = \gamma = 90°$. And yet when one examines the crystal structure, i.e., the atomic arrangements within a unit cell, the four 3-fold axes are missing. In this case, it is an apparent metric relationship that arises accidentally. An example of this is provided by the material lead zirconate, $PbZrO_3$, where measurements of the unit cell were originally found to suggest $a = b$ $\alpha = \beta = \gamma = 90°$, which is consistent with tetragonal symmetry. However, when one examines the arrangement of atoms in the crystal it is obvious that there is no 4-fold axis of symmetry present. Very careful

Table 1.4. The seven crystal systems and 32 geometric crystal classes.

Crystal System	Schoenflies	International	Axes restrictions
Triclinic	C_1	1	-
	$S_2(C_i)$	$\bar{1}$	
Monoclinic	C_2	2	$\alpha = \gamma = 90°$
	$C_{1h}(C_S)$	m	
	C_{2h}	2/m	
Orthorhombic	$D_2(V)$	222	$\alpha = \beta = \gamma = 90°$
	C_{2v}	mm2	
	$D_{2h}(V_h)$	mmm	
Tetragonal	C_4	4	$a = b$; $\alpha = \gamma = 90°$
	S_4	$\bar{4}$	
	C_{4h}	4/m	
	D_4	422	
	C_{4v}	4mm	
	$D_{2d}(V_d)$	$\bar{4}$2m	
	D_{4h}	4/mmm	
Trigonal	C_3	3	$a = b$; $\alpha = \beta = 90°$ $\gamma = 120°$
	$S_6(C_{3i})$	$\bar{3}$	
	D_3	32	
	C_{3v}	3m	
	D_{3d}	$\bar{3}$m	
Hexagonal	C_6	6	$a = b$; $\alpha = \beta = 90°$ $\gamma = 120°$
	C_{3h}	$\bar{6}$	
	C_{6h}	6/m	
	D_6	622	
	C_{6v}	6mm	
	D_{3h}	$\bar{6}$m2	
	D_{6h}	6/mmm	
Cubic	T	23	$a = b = c$ $\alpha = \beta = \gamma = 90°$
	T_h	m$\bar{3}$	
	O	432	
	T_d	$\bar{4}$3m	
	O_h	m$\bar{3}$m	

measurements carried out later in fact showed that the angle $\gamma = 89.70°$ and not exactly 90° as would be required for tetragonal symmetry.

The limit in the number of crystal systems arises because only 1, 2, 3, 4 and 6-fold operations are possible in a conventional three-dimensional lattice. This can be seen by reference to figure 1.18.

In the diagram four lattice points are marked. The points at A and B are separated by a lattice translation t, whereas the points C and D are separated by an unknown number of lattice translations nt. The line AC represents the effect of rotating AB clockwise about A through the angle α, and BD, the corresponding

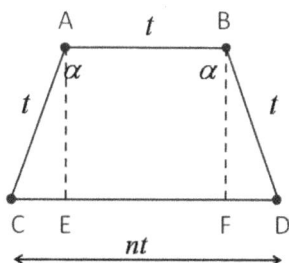

Figure 1.18. Limitations on rotational symmetry for a lattice.

effect of rotating anticlockwise about B. We see from this that the distance EF = t. Now by considering triangles ACE and BDF, we find that CE = DF = $t \sin \alpha$, so that

$$nt = CD = t + 2t \sin \alpha \qquad (1.16)$$

Therefore

$$\sin \alpha = \frac{n-1}{2} \qquad (1.17)$$

from which we obtain the following values for the rotation angle α:

n	α
0	120°
1	0°, 180°
2	60°
3	90°

You can see from this that a 5-fold rotation ($\alpha = 72°$) is not possible. Another way of thinking about this is in terms of tiling a plane with regular polygons. If you try this with 5-fold and 7-fold polygons you will find that however hard you try stacking them together they will always leave spaces. British 20 and 50 pence coins have 7 sides, thus making it easier to pick them up from a counter.

Bravais lattices

In 1854 Auguste Bravais (1811–63)[1] considered whether there was a limit to the number of *unique* types of lattice that could be defined. For three-dimensional lattices, he determined that there were 14 in all.

---

[1] I am always amused when I see Bravais referred to as M Bravais in some books, the authors apparently unaware that the M stands for Monsieur!

Figure 1.19. The 14 Bravais lattices. The black spheres indicate lattice points, and should not be mistaken for atoms!

All 14 are shown in figure 1.19. Thus, for example, there are three Bravais lattices that are compatible with cubic symmetry, defined by the presence of four 3-fold axes, P, I and F. A one-face-centred unit cell would not be cubic (even though metrically it might accidentally look like a cube) because the 3-fold axes along the four body-diagonals of the cube would want to make each face equivalent. One way to derive the 14 Bravais lattices is to search for the smallest unit cell possible in any lattice that at the same time shows the true symmetry of the lattice. Thus, for example, in the tetragonal system there are two Bravais lattices, labelled here as P and I.

So, what about the possibility of C and F? As can be seen (figure 1.20), by redefining a smaller unit cell through rotation by 45° the 4-fold symmetry necessary for a tetragonal lattice is retained and then $C = P$ and $F = I$, and as a result, tetragonal C and F lattices are not uniquely different. Which one chooses is a matter of personal preference. Thus, one could choose to specify axes such that in, say, the triclinic system the unit cell is all-face-centred rather than primitive This would make a unit cell four times as big as is necessary (containing four lattice points rather than one) but there could be reasons why in a particular case one would prefer this larger

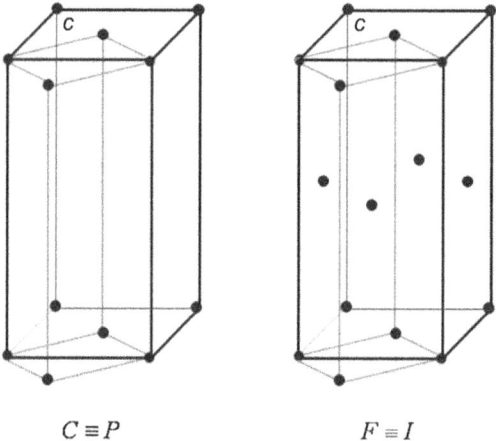

Figure 1.20. Tetragonal system: showing equivalence of C and P, and I and F.

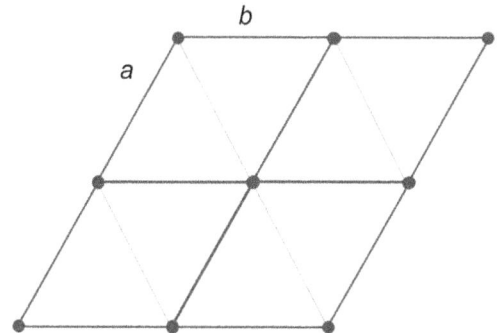

Figure 1.21. A primitive hexagonal lattice viewed on (0001).

unit cell: for instance, it might make it easier to compare two different phases of a material.

Hexagonal, trigonal (and rhombohedral)

You will notice in figure 1.19 that in addition to a hexagonal P lattice there is a trigonal lattice given the symbol R. The three terms hexagonal, trigonal and rhombohedral often cause confusion and need special attention. Let us first of all consider a hexagonal primitive lattice. This is shown in figure 1.21. The figure shows four unit cells, whence it can be seen that each lattice point is surrounded by six lattice points to form a hexagon[2]. Suppose now we draw the equivalent diagram for a trigonal lattice. The strange thing is that when we do this it looks exactly the same! So,

---

[2] Physicists often call this a triangular lattice, in my view incorrectly. You cannot make a lattice consisting of triangles, because you end up with two triangles pointing in opposite directions, i.e., the triangle does not form a unit cell.

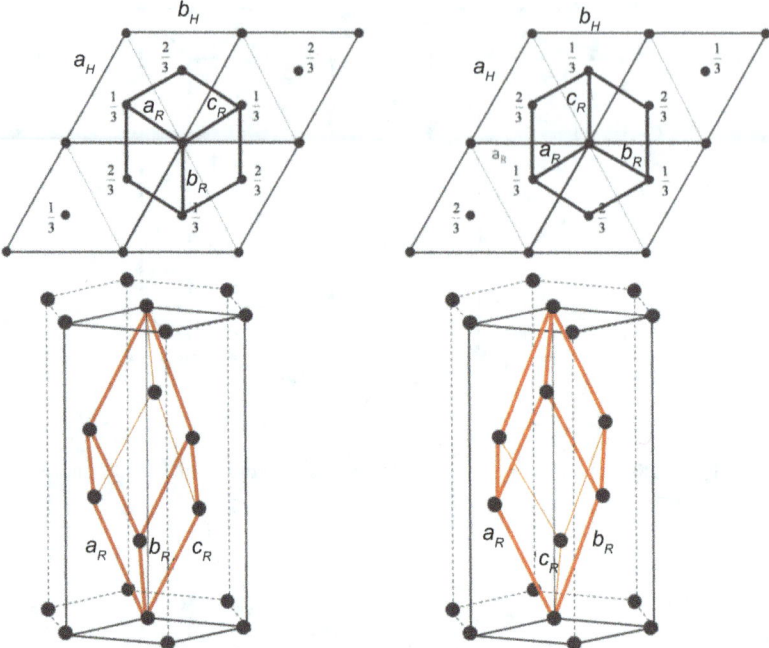

Figure 1.22. Trigonal centering with primitive rhombohedral unit cell added. Left is the obverse setting and right is the reverse setting.

does this mean that in fact the trigonal and hexagonal systems are equivalent? This has been the object of much debate in the past, because as far as the lattices are concerned there is no distinct trigonal system. However, when one examines the atomic positions in the crystal structure itself it may turn out that the arrangement of atoms shows just a single 3-fold rather than 6-fold symmetry, thus justifying the term trigonal.

Now consider the left-hand side of figure 1.22. This time two extra lattice points have been added in each unit cell at heights $c/3$ and $2c/3$. Notice now that a 3-fold symmetry is present, for example relating the lattice points at $c/3$. We can therefore think of this as a centred trigonal lattice. Furthermore, by joining up the points a smaller unit cell with axes a_R, b_R and c_R can be constructed. This unit cell is primitive and contains a single 3-fold axis. It also has the property that

$$a_R = b_R = c_R \quad \alpha = \beta = \gamma \tag{1.18}$$

When the lattice is described in this way it known as *rhombohedral* and is given the symbol R. This can be seen in figure 1.19. In the International Tables for Crystallography, where all aspects of crystal symmetry are carefully defined and explained, the convention is to include the rhombohedral symmetry as part of the trigonal system, so that one can freely choose between centred trigonal and rhombohedral geometry.

Be aware of another complexity. The drawing on the left-hand side of figure 1.22 is known as the obverse setting, whereas the one on the right is the so-called reverse setting. It is the normal convention to use the obverse setting in describing rhombohedral systems, although you may encounter the reverse setting from time to time.

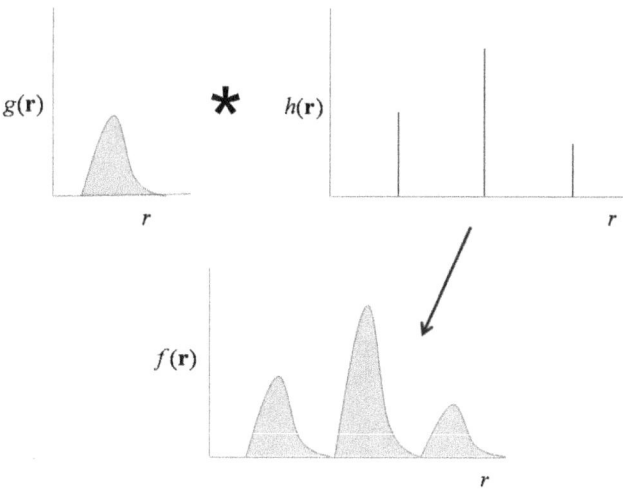

Figure 1.23. An example of the convolution of two functions.

1.6 Crystal structures

Some studies pursued by the writer as to the nature of molecules have led him to believe that in the atom-groupings which modern chemistry reveals to us, the several atoms occupy distinct portions of space and do not lose their individuality.
William Barlow, 1883 paper in which he predicted the structure of alkali halides

We now put together the ideas discussed hitherto in order to explain what is really meant by the term crystal.

Definition. *In general a crystal is a solid in which a group of atoms or molecules adopt on <u>average</u> an ordered arrangement in space.*

Actually, in a real crystal the atoms are in constant vibration and there may be all sorts of defects and impurities present, and so our definition invoking the concept of order refers to the average positions of the atoms or molecules. Note that this definition refers to atoms and molecules and not lattice points. Once again, please do not confuse the lattice with the structure. Now, one of the simplest ways to define a crystal and its structure is through use of the mathematical operation of convolution.

Convolution

Consider two functions $g(\mathbf{r})$ and $h(\mathbf{r})$ spanning a space given by the vector $\mathbf{r}$. The convolution of these two functions $f(\mathbf{r})$ is given by

$$f(\mathbf{r}) = g(\mathbf{r}) * h(\mathbf{r}) = \int_{-\infty}^{\infty} g(\mathbf{r}')h(\mathbf{r} - \mathbf{r}')d\mathbf{r}' \tag{1.19}$$

The effect of this operation is to slide one function over the other whilst integrating. Figure 1.23 illustrates this process for two arbitrarily chosen functions. It can be seen that here, where one of the functions consists of three delta functions, the effect is to repeat the first function in sympathy with the delta functions.

Convolution applied to crystals

Suppose we have a collection of atoms described by a function $B(\mathbf{r})$: often in physics texts this is referred to as the 'basis' (a term by the way that is not used by crystallographers, where basis refers to the set of basis vectors defining the axes of a unit cell). In order to form a crystal structure, we need to repeat this basis according to the translational symmetry of the lattice. The lattice function $L(\mathbf{r})$ can be defined by

$$L(\mathbf{r}) = \sum_{uvw} \delta(\mathbf{r} - \mathbf{r}_{uvw}) \qquad (1.20)$$

and then the crystal structure $C(\mathbf{r})$ is given by

$$C(\mathbf{r}) = L(\mathbf{r}) * B(\mathbf{r}) \qquad (1.21)$$

In figure 1.12 the function $B(\mathbf{r})$ was chosen to be represented by an elephant! Now convolute this with the lattice in figure 1.13 and the result will be the 'crystal' structure, in this case consisting of a regular array of elephants. Of course, this is not a real crystal structure but it does illustrate how the lattice acts as a template instructing how the basis is to be repeated.

Examples of crystal structures

The concept of a basis convoluted with a lattice is a useful and convenient way to describe relatively simple crystal structures. Here we shall consider a few examples.

1. Copper Cu
    The crystal structure of the metal copper (figure 1.24) is very simple and can be summarised thus:

Crystal system	Cubic	$a = 3.610$ Å
Lattice	F	(0, 0, 0) (½, ½, 0) (½, 0, ½) (0, ½, ½)
Basis	Cu	(0, 0, 0)
Number of atoms in unit cell	Z	4
Fractional coordinates	Cu	(0, 0, 0) (½, ½, 0) (½, 0, ½) (0, ½, ½)

This illustrates why some people become confused between lattices and structures. This structure has a basis consisting of one atom, so that when we construct the structure we see that there is a single atom for each lattice

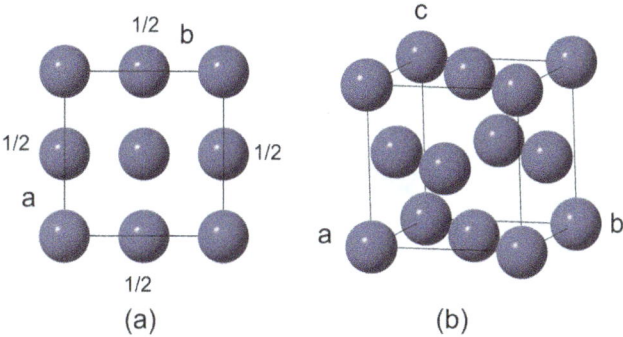

Figure 1.24. Copper crystal structure (a) (001) projection (b) perspective view.

point. The result is a diagram that closely resembles the equivalent drawing for the lattice, as seen in figure 1.19: but remember in the picture of the structure we have atoms, not points!

2. Molybdenum Mo

Once again this structure (figure 1.25) has a basis consisting of a single atom and a body-centred lattice, and so there is one atom per lattice point.

Crystal system	Cubic	$a = 3.150$ Å
Lattice	I	(0, 0, 0) (½, ½, ½)
Basis	Mo	(0, 0, 0)
Number of atoms in unit cell	Z	2
Fractional coordinates	Mo	(0, 0, 0) (½, ½, ½)

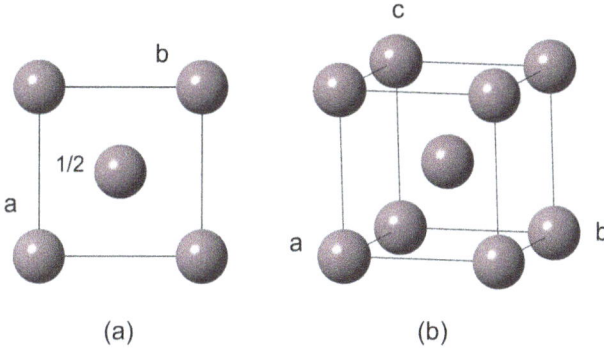

Figure 1.25. Molybdenum crystal structure (a) (001) projection (b) perspective view.

Note that the crystal structure of iron Fe has the same atomic arrangement.

3. Sodium Chloride NaCl

Crystal system		Cubic	$a = 5.628$ Å
Lattice		F	(0, 0, 0) (½, ½, 0) (½, 0, ½) (0, ½, ½)
Basis		Na	(0, 0, 0)
		Cl	(½, 0, 0)
Number of atoms in unit cell		Z	8
Fractional coordinates		Na	(0, 0, 0) (½, ½, 0) (½, 0, ½) (0, ½, ½)
		Cl	(½, 0, 0) (0, ½, 0) (0, 0, ½) (½, ½, ½)

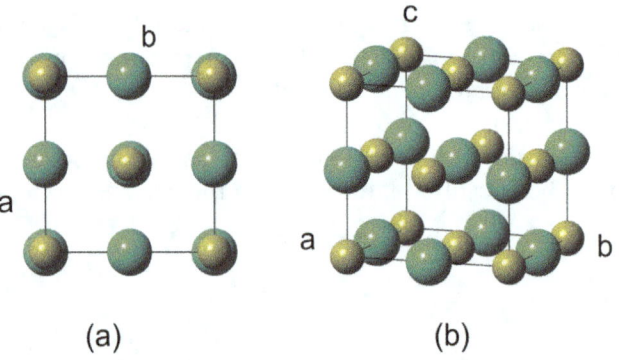

Figure 1.26. The sodium chloride crystal structure (a) (001) projection (b) perspective drawing. Na: yellow, Cl: green.

The structure of the alkali halides, as exemplified by NaCl (figure 1.26), consists of an alternating arrangement of cations and anions, as shown in figure 1.26. This structure is an example of one with a basis consisting of two atoms. The result of convoluting the four lattice points for the *F* unit cell with the two atoms in the basis is to create eight atoms in the unit cell.

4. Cesium Chloride CsCl

This structure (figure 1.27) is what I call 'an old chestnut' as it has been used in countless condensed matter physics examination questions.

Crystal system		Cubic	$a = 4.123$ Å
Lattice		P	(0, 0, 0)
Basis		Cs	(0, 0, 0)
		Cl	(½, ½, ½)
Number of atoms in unit cell		Z	2
Fractional coordinates		Na	(0, 0, 0)
		Cl	(½, ½, ½)

A Journey into Reciprocal Space

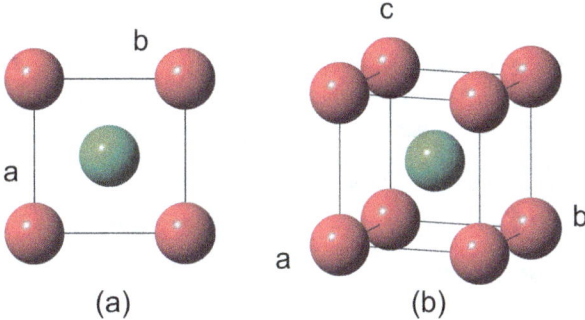

Figure 1.27. Cesium chloride structure (a) (001) projection (b) perspective view Cs: pink Cl: green.

At first sight (figure 1.27), this structure looks like that of molybdenum, but the difference here is that the atom at the centre of the unit cell is not the same as the one at the corners. Therefore, it is a mistake to say that the lattice is body-centred, as implied misleadingly in some books, but it is primitive P with a basis consisting of two atoms.

The crystal structure of β-brass (CuZn) at low temperature is similar. In this structure, as temperature is increased the Cu and Zn atoms tend to hop onto each other's sites, until at a certain temperature each of the two atomic sites has on average ½(Cu + Zn).

The structure, when averaged over all unit cells, is then effectively body-centred cubic, as in molybdenum. This is an example of an order–disorder phase transition.

5. Diamond C

Here is another old favourite and possibly the most important crystal structure (figure 1.28) for physicists.

Crystal system	Cubic	$a = 5.646$ Å
Lattice	F	(0, 0, 0) (½, ½, 0) (½, 0, ½) (0, ½, ½)
Basis	C	(0, 0, 0)
	C	(¼, ¼, ¼)
Number of atoms in unit cell	Z	8
Fractional coordinates	C1	(0, 0, 0) (½, ½, 0) (½, 0, ½) (0, ½, ½)
	C2	(¼, ¼, ¼) (¾, ¾, ¼) (¾, ¼, ¾) (¼, ¾, ¾)

This again is a structure with two atoms in the basis, C1 and C2. Notice how the convolution of the basis with the lattice results again in eight atoms in the unit cell, but with a structure that is totally different from that of NaCl. Whilst these two crystals have the same type of lattice, it is the differences in the bases that create the different crystal structures. This is why phrases like

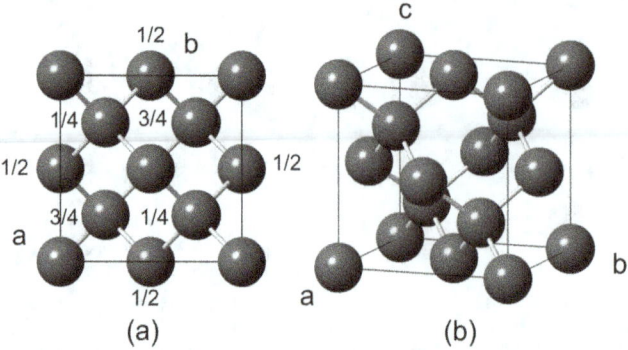

Figure 1.28. Diamond/silicon/germanium structure (a) (001) projection (b) perspective view.

'the diamond lattice' are confusing, because the diamond lattice means face-centred cubic: what is meant is 'the diamond crystal structure'. Sorry to labour the point but it is important to understand this distinction between lattice and structure. The same basic structure type is adopted by silicon and germanium. This structure type is centrosymmetric: see if you can work out where the centre of inversion lies.

6. Zinc Sulfide ZnS

This structure type, otherwise known as the zinc blende or sphalerite structure (figure 1.29), and adopted by many compounds such as GaAs, InSb etc, looks at first sight to be the same as that of diamond.

Crystal system	Cubic	$a = 5.420$ Å
Lattice	F	(0, 0, 0) (½, ½, 0) (½, 0, ½) (0, ½, ½)
Basis	Zn	(0, 0, 0)
	S	(¼, ¼, ¼)
Number of atoms in unit cell	Z	8
Fractional coordinates	Zn	(0, 0, 0) (½, ½, 0) (½, 0, ½) (0, ½, ½)
	S	(¼, ¼, ¼) (¾, ¾, ¼) (¾, ¼, ¾) (¼, ¾, ¾)

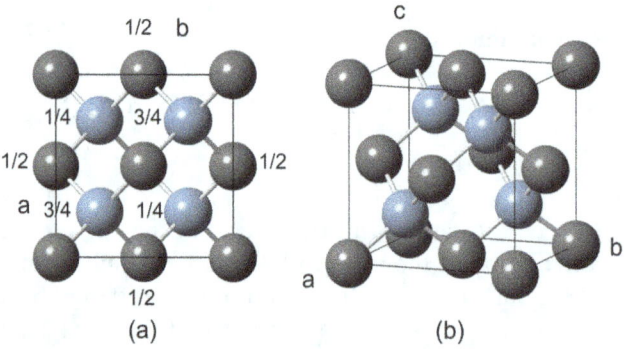

Figure 1.29. Zinc blende structure (a) (001) projection (b) perspective view Zn: grey S: blue.

However, the difference here is that the two atoms in the basis are not the same element. This structure type is non-centrosymmetric and so this material can exhibit a polar property such as piezoelectricity (but not pyroelectricity, as it has no unique polar axis).

7. Perovskite

This structure (figure 1.30) is one of my favourites, since I have spent most of my research life studying these compounds. It is adopted by many compounds of general formula ABX_3, where A and B are cations and X is an anion, usually oxygen. The structure described below is the usual high-temperature structure of many perovskites such as that found in $SrTiO_3$.

Crystal system	Cubic	$a = 3.91$ Å
Lattice	P	(0, 0, 0)
Basis	Sr	(½, ½, ½)
	Ti	(0, 0, 0)
	O	(½, 0, 0)
Number of atoms in unit cell	Z	5
Fractional coordinates	Sr	(½, ½, ½)
	Ti	(0, 0, 0)
	O	(½, 0, 0) (0, ½, 0) (0, 0, ½)

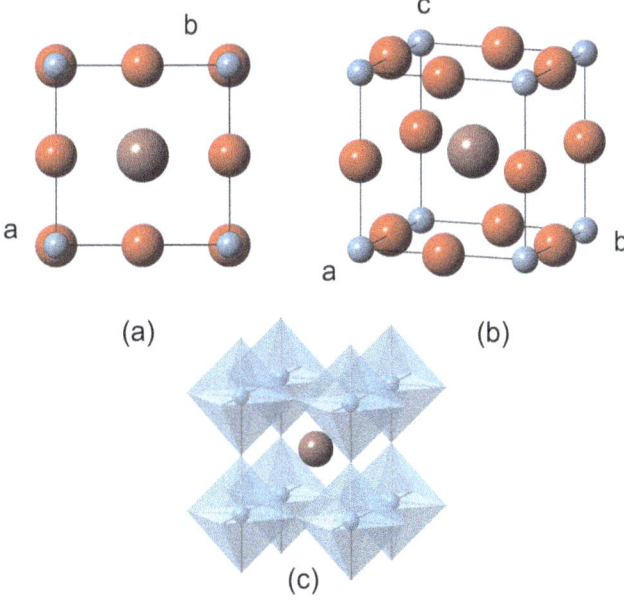

Figure 1.30. The $SrTiO_3$ crystal structure (a) (001) projection (b) perspective drawing (c) showing oxygen octahedra. Sr: maroon, Ti: blue, O: red.

Figure 1.30(a) and (b) show the usual drawings of the structure while figure 1.30(c) shows that when the O–O bonds are drawn in the result is a framework of corner-linked octahedra (in the figure the oxygen atoms are not shown but occur at the vertices of the octahedra). Drawings of structures that emphasize polyhedral coordination in this way are often to be found in the literature.

8. Wurtzite ZnS

Just to show that not everything is cubic, consider the wurtzite structure (figure 1.31). This is another form of zinc sulfide. The crystal system in this case is hexagonal.

Crystal system	Hexagonal	$a = 3.81$ Å. $c = 6.23$ Å
Lattice	P	(0, 0, 0)
Basis	Zn1	(1/3, 2/3, 0)
	Zn2	(2/3, 1/3, ½)
	S1	(1/3, 2/3, 0.375)
	S2	(2/3, 1/3, ½+0.375)
Number of atoms in unit cell	Z	4
Fractional coordinates	Zn	(1/3, 2/3, 0) (2/3, 1/3, ½)
	S	(1/3, 2/3, 0.375) (2/3, 1/3, ½+0.375)

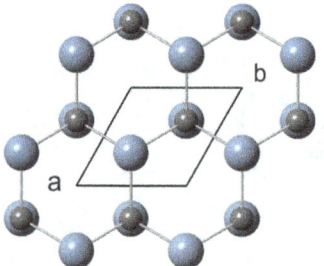

Figure 1.31. The wurtzite crystal structure: (0001) projection. Zn: grey, S: blue.

The lattice is primitive and there are four atoms in the basis and therefore in the unit cell. Notice that in this structure there is just one freely variable parameter for the S atomic coordinates. Having fixed the Zn positions, the z-coordinate of the S atoms obeys the relationship

$$z(S1) = \frac{1}{2} + z(S2) \tag{1.22}$$

Definition. *The ability for a chemical entity, in this case ZnS, to exist in different crystal structures is called polymorphism.*

1.7 Space groups

The concept of a lattice and a basis, as used sometimes in condensed matter textbooks, is not used by crystallographers, who use instead space group descriptions of crystal structures. A full description of space groups lies outside the scope of this book and the reader should consult the International Tables for Crystallography, Volume A (ITA) [5] or the book by Burns and Glazer [7]. I regard the ITA as one of the greatest books produced in the 20th century! Here I shall only give a brief explanation. I must first of all discuss a useful concept that we shall need to use, namely the *asymmetric unit*.

Definition. *The asymmetric unit is a part of space that generates the whole of space when all the symmetry operations of a space group are applied.*

Thus, in the example above of diamond, the asymmetric unit contains a single atom of carbon, as opposed to the basis consisting of two carbon atoms. A crystallographer would describe the crystal structure of diamond by specifying the space group by the symbol $Fd\bar{3}m$ with a carbon atom at (0, 0, 0). Notice that we now only have to specify the position of one carbon atom, instead of two, but let the space group with its symmetry generate the remaining seven atoms in the unit cell, thus a saving of 50% of the information needed. Symmetry allows us to specify fewer parameters than otherwise.

Space groups are constructed in principle by combining the 32 point groups with the 14 Bravais lattices. Thus, for instance the point group $mm2$ can be combined with body-centring to give the space group $Imm2$, or say $2/m$ with a primitive lattice to give $P2/m$.

Figure 1.32 shows how space group information is given in the ITA for space group $P2/m$. Reading from left to right at the top we see the crystal system (monoclinic) followed by the point group ($2/m$). Then the so-called full symbol P $112/m$ appears: the 1's just mean that there are no symmetry elements along the a and b axes, while the $2/m$ which appears in the third place of the symbol means that there is a 2-fold axis along c with a perpendicular mirror plane. On the right is the space group symbol in International notation $P2/m$ and underneath the Schoenflies

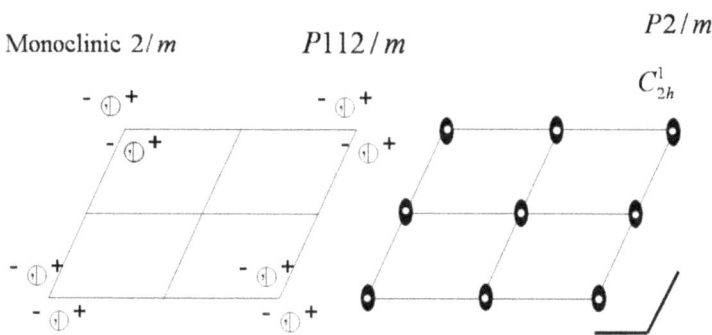

Figure 1.32. Example of space group $P2/m$ *(1st Setting)*.

symbol C_{2h}^1. The Schoenflies symbol is not much used for space groups because differentiation between space groups with the same point symmetry is by a numerical superscript and the only way to know what this means is to look it up in a book. The International notation, on the other hand, is sufficient for a competent person to derive all the space group information. Below the summary information are two diagrams, one on the left showing the formation of general equivalent positions by the space group symmetry and one on the right the locations of the symmetry elements. Each diagram shows a unit cell which by convention is drawn with the a-axis down, the b-axis to the right and the c-axis out of the page, thus placing the origin of the unit cell at the top left of each unit cell drawing.

Space groups formed in this way are known as *symmorphic* space groups, of which there are 73 in total, and are defined by the Seitz operator:

$$\{R|\mathbf{t}_n\}\mathbf{r} = R\mathbf{r} + \mathbf{t}_n \tag{1.23}$$

R represents a point group symmetry operator and after applying it to the position vector $\mathbf{r}$, lattice translations $\mathbf{t}_n$ are added. The Seitz operators form a so-called *real affine group*. However, once one combines lattices with point symmetry it is found that new types of symmetry operation are required. The operations include a translation that is less than a full lattice translation and given by the Seitz operator as

$$\{R|\mathbf{v} + \mathbf{t}_n\}\mathbf{r} = R\mathbf{r} + \mathbf{v} + \mathbf{t}_n \tag{1.24}$$

or within a single unit cell we can omit the $\mathbf{t}_n$ for convenience

$$\{R|\mathbf{v}\}\mathbf{r} = R\mathbf{r} + \mathbf{v} \tag{1.25}$$

Space groups that are defined in this way are known as *non-symmorphic* space groups. These contain the so-called screw and glide operations. In total, there are 157 of these making a total of 230 three-dimensional space group types.

Screw axes

Screw axes consist of a rotation plus a fractional translation along the rotation axis.
The 2-fold screw operator about [001], symbol, 2_1, (figure 1.33) can be written as

$$\{2_{001}|(0, 0, 1/2\} \tag{1.26}$$

Note that applying this operator twice gives

$$\{1|(0, 0, 1\} \equiv \{1|(0, 0, 0)\} \tag{1.27}$$

i.e. equivalent to the identity operation within the original unit cell.

Similarly, consider the four-fold screw axes 4_1, 4_2 and 4_3 directed along [001]. Figure 1.34 shows diagrams for space groups $P4_1$ and $P4_3$. For fun, the space groups are illustrated by pictures of an aircraft undergoing barrel-rolls towards us to the left and to the right, respectively (naturally, in a real crystal, instead of an aircraft, one would have a molecule or group of atoms, but the principle is nevertheless true).

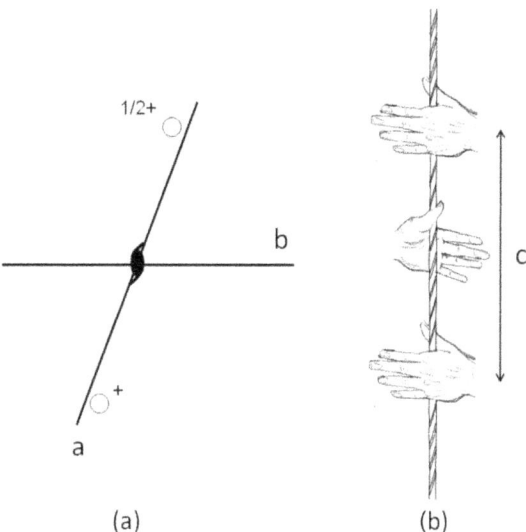

Figure 1.33. A $2_1[001]$ screw axis. (a) Stereographic projection (b) An example showing hands related by a 2-fold screw axis.

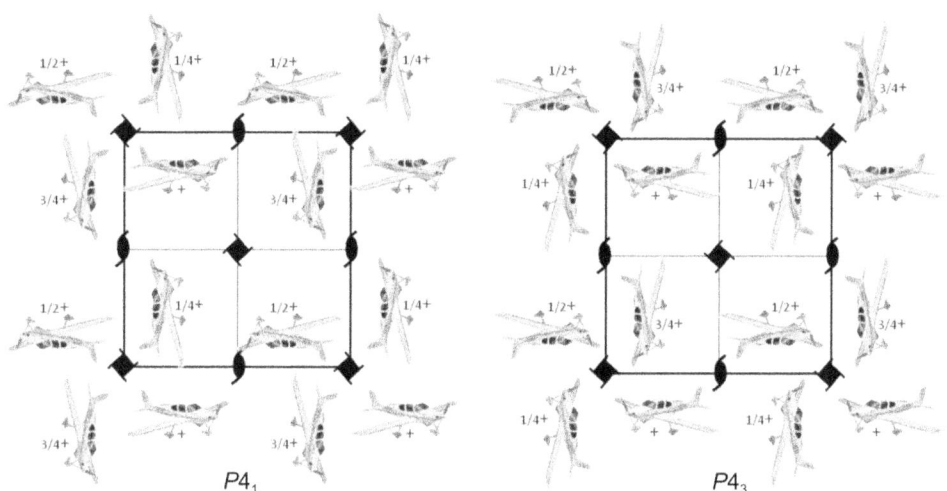

Figure 1.34. Examples of space groups $P4_1$ and $P4_3$.

Notice that the combination of 4_1 and 4_3 axes with a primitive tetragonal lattice gives rise to 2-fold screw axes mid-way between the 4-fold screw axes. It can also be seen that 4_1 and 4_3 describe helices of opposite chirality. Thus, these two space groups are mirror images of each other, and as a result it has been argued in the past that they should not really be considered to be separate space groups. The International Tables nevertheless does include them separately. $P4_1$ and $P4_3$ are examples of what are termed *enantiomorphic* space group types. There are 11 such enantiomorphic space group types in three dimensions.

The Seitz operators for 4_1 and 4_3 operations about the c-axis are

$$\{4_{001}|(0, 0, \tfrac{1}{4})\} \quad \text{and} \quad \{4_{001}|(0, 0, \tfrac{3}{4})\} \tag{1.28}$$

Glide planes

These involve a reflection plus a fractional translation. Thus, for instance a reflection across the (010) plane plus a translation of halfway along c is given by

$$\left\{m_{010}|(0, 0, \tfrac{1}{2})\right\} \tag{1.29}$$

m_{010} refers to a reflection across the plane perpendicular to the b-axis. This is known as an axial c-glide (a and b-glides involve half translations along a and b, respectively). Figure 1.35 shows space group Pa (or full symbol $P1a1$) using an aircraft as a general motif! Note that the two aircraft within the unit cell are mirror images of each other and are not related by rotational symmetry: look carefully at the directions of the propeller blades in each case!

There are other types of glide planes possible (diagonal, diamond and double glides). Together with the screw axes it is found that there are 230 space group types in three dimensions. For further information, I recommend consulting the various books devoted to space groups. Below I list the crystal structures from above using space groups.

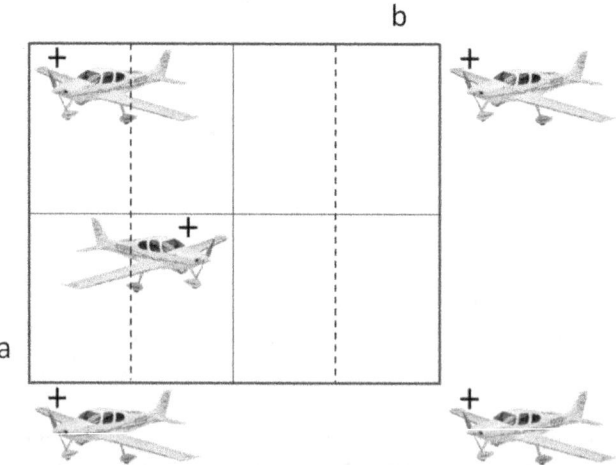

Figure 1.35. Example of an a-glide in space group P $1a1$ (Pa)—2nd setting monoclinic.

Cu	$Fm\bar{3}m$	Cu:	0, 0, 0
Mo	$Im\bar{3}m$	Mo:	0, 0, 0
NaCl	$Fm\bar{3}m$	Na:	0, 0, 0
		Cl:	½, 0, 0
CsCl	$Pm\bar{3}m$	Cs:	0, 0, 0
		Cl:	½, ½, ½
Diamond	$Fd\bar{3}m$	C:	0, 0, 0
Zinc blende	$F\bar{4}3m$	Zn:	0, 0, 0
		S:	¼, ¼, ¼
SrTiO$_3$	$Pm\bar{3}m$	Sr:	½, ½, ½
		Ti:	0, 0, 0
		O:	½, 0, 0
Wurtzite	$P6_2mc$	Zn:	1/3, 2/3, 0
		S:	1/3, 2/3, 0.375

References

[1] Hammond C 2015 *The Basics of Crystallography and Diffraction* (Oxford: Oxford University Press)
[2] Giacovazzo C, Monaco H L, Artioli G, Viterbo D, Milanesio M, Gilli G, Gilli P, Zanotti G, Ferraris G and Catti M 2011 *Fundamentals of Crystallography* (Oxford: Oxford University Press)
[3] Glazer A M 2016 *Crystallography: A Very Short Introduction* (Oxford: Oxford University Press)
[4] Online Dictionary of Crystallography [Online]. Available: http://reference.iucr.org/dictionary/Main_Page
[5] Aroyo M I 2016 International Tables for Crystallography *vol* A (Chester: International Union of Crystallography)
[6] Donnay J D H and Le Page Y 1978 The vicissitudes of the low-quartz crystal setting or the pitfalls of enantiomorphism *Acta Crystallogr. Sect.* A **34** 584–94
[7] Burns G and Glazer A M 2013 *Space Groups for Solid State Scientists* 3rd edn (Cambridge, MA: Academic Press)

IOP Concise Physics

A Journey into Reciprocal Space
A crystallographer's perspective
A M Glazer

Chapter 2

The reciprocal lattice

Doc Glazer he plays with x-rays sir
Doc Glazer he plays with x-rays sir
The reciprocal lattice
Who on earth knows what that is
Oh k-space, yes that's OK—sir!

(P A Thomas: to the tune of 'It Ain't Necessarily So' from an undergraduate party at Jesus College Oxford)

Brief history

M L Frankenheim and A Bravais in the mid-19th century showed that in direct space symmetry placed a restriction on the number of unique descriptions of a lattice (14 in three dimensions, for example). Bravais in 1850 also considered another type of lattice, called the *polar lattice* [1] obtained by forming points on vectors normal to the lattice planes. The repeat distances between the points were proportional to the reciprocal of the interplanar distances d_{hkl} of the direct lattice according to

$$\frac{V^{2/3}}{d_{hkl}} \qquad (2.1)$$

Thus, the distances in this polar lattice were expressed in Å units. J W Gibbs in the 1880s used this concept in his lectures, but with the repeat distances now in the form

$$\frac{1}{d_{hkl}} \qquad (2.2)$$

and so the units of this lattice were expressed in $Å^{-1}$. However, it was P P Ewald, who, in or around 1913, used this definition to explain the arrangement of spots seen when x-rays are diffracted by a crystal. This is the form used by crystallographers to the present day in interpreting their diffraction patterns from crystals.

2.1 Definition

To explain what is meant by the reciprocal lattice I will first define it through mathematical statements. Let us start by defining its axes in much the same way as was done for direct lattices. Thus, instead of the axes and angles $a, b, c, \alpha, \beta, \gamma$ defining the crystallographic basis of the direct lattice, we define a corresponding set denoted by $a^*, b^*, c^*, \alpha^*, \beta^*, \gamma^*$. The reciprocal axis vectors $\mathbf{a}^*$, $\mathbf{b}^*$ and $\mathbf{c}^*$ are then given by the following equations

$$\mathbf{a}^* = 2\pi \frac{\mathbf{b} \times \mathbf{c}}{\mathbf{a} \cdot \mathbf{b} \times \mathbf{c}} = 2\pi \frac{\mathbf{b} \times \mathbf{c}}{V}$$
$$\mathbf{b}^* = 2\pi \frac{\mathbf{c} \times \mathbf{a}}{\mathbf{a} \cdot \mathbf{b} \times \mathbf{c}} = 2\pi \frac{\mathbf{c} \times \mathbf{a}}{V} \qquad (2.3)$$
$$\mathbf{c}^* = 2\pi \frac{\mathbf{a} \times \mathbf{b}}{\mathbf{a} \cdot \mathbf{b} \times \mathbf{c}} = 2\pi \frac{\mathbf{a} \times \mathbf{b}}{V}$$

V is the volume of the corresponding unit cell in direct space. Using scalar products of the reciprocal axis vectors, it can be shown that the reciprocal angles α^*, β^* and γ^* are related to the real angles α, β and γ by

$$\cos \alpha^* = \frac{\cos \beta \cos \gamma - \cos \alpha}{\sin \beta \sin \gamma}$$
$$\cos \beta^* = \frac{\cos \gamma \cos \alpha - \cos \beta}{\sin \gamma \sin \alpha} \qquad (2.4)$$
$$\cos \gamma^* = \frac{\cos \alpha \cos \beta - \cos \gamma}{\sin \alpha \sin \beta}$$

Before going on, the factor 2π that appears in equation (2.3) needs to be explained. This is simply a constant of proportionality and could in fact be replaced by any value we choose. The factor of 2π is commonly used by physicists [2–4], who are normally interested in the propagation of waves in solids. The modulus of the wave-vector $\mathbf{k}$ of a wave is related to the wavelength λ by:

$$k = \frac{2\pi}{\lambda} \qquad (2.5)$$

with wave momentum given by $\hbar k$. On the other hand, crystallographers usually replace it by 1. As this book is addressed primarily to non-crystallographers, I shall retain the factor of 2π throughout.

Consider now the scalar product

$$\mathbf{a} \cdot \mathbf{a}^* = 2\pi \frac{\mathbf{a} \cdot \mathbf{b} \times \mathbf{c}}{\mathbf{a} \cdot \mathbf{b} \times \mathbf{c}} = 2\pi \qquad (2.6)$$

and also

$$\mathbf{a} \cdot \mathbf{b}^* = 2\pi \frac{\mathbf{a} \cdot \mathbf{c} \times \mathbf{a}}{V} = 2\pi \frac{\mathbf{a} \cdot \mathbf{n} \sin \beta}{V} = 0 \tag{2.7}$$

where the vector **n** is perpendicular to the basis vectors **a** and **c**.

So, one of the consequences of the definitions of the reciprocal axis vectors is that the following relationships between real and reciprocal axes always apply

$$\begin{aligned}
\mathbf{a} \cdot \mathbf{a}^* &= \mathbf{b} \cdot \mathbf{b}^* = \mathbf{c} \cdot \mathbf{c}^* = 2\pi \\
\mathbf{a} \cdot \mathbf{b}^* &= \mathbf{a} \cdot \mathbf{c}^* = \mathbf{b} \cdot \mathbf{c}^* = \mathbf{b} \cdot \mathbf{a}^* = \mathbf{c} \cdot \mathbf{b}^* = \mathbf{c} \cdot \mathbf{a}^* = 0 \\
\mathbf{a}^* \cdot \mathbf{a}^* &= a^{*2}; \; \mathbf{a}^* \cdot \mathbf{b}^* = a^*b^* \cos \gamma^*, \ldots\ldots
\end{aligned} \tag{2.8}$$

Thus, we see that any real axis, say **a**, is always perpendicular to the reciprocal **b*** and **c*** axes, with similar relationships between all the other axes. Furthermore, note that in general, any real axis is only parallel to its related reciprocal axis when the geometry of the unit cell is described by orthogonal axes.

2.2 Construction

We can now use the above definitions to construct a drawing of the reciprocal lattice starting from a real lattice. Consider first a unit cell of a real lattice with axes **a** and **b** as shown in figure 2.1. This cell has deliberately been chosen to be oblique, with the angle between the axes denoted by γ, to illustrate the general case more easily.

In figure 2.1 the separation d_{100} of the (100) set of parallel planes is marked (in grey) in the real unit cell. Starting with a chosen origin for this unit cell, the red line drawn perpendicular to the (100) planes denotes the **a*** axis, and a point is added, labelled 100, at a distance from the origin given by $g_{100} = 2\pi/d_{100}$. An additional point labelled 200 is added at twice this distance, thus corresponding in real space to

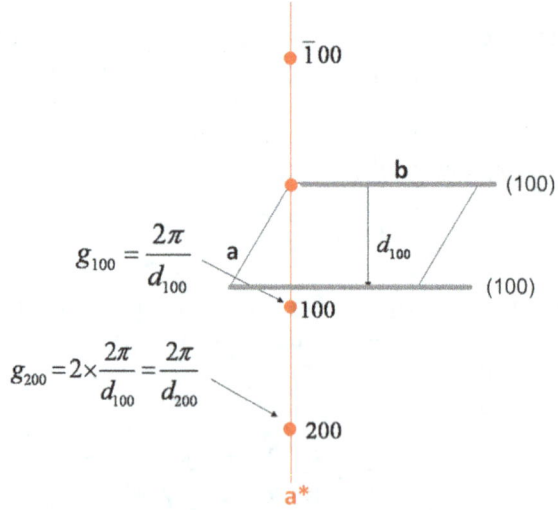

Figure 2.1. Construction of 100, 200 ... reciprocal lattice points.

(200) planes i.e. planes of half the spacing between the (100) planes. Note therefore that as the planes get closer together in real space, the corresponding reciprocal lattice points become further away from the origin. You can begin to see now why we use the term *reciprocal space*.

Proceeding in a similar way we mark a point on the −**a** axis, denoted $\bar{1}00$ (note the bar placed above the 1). In figure 2.2 the same procedure has been used for the set of planes (010), (020), (030), etc, and their negative values. It can be seen that the respective points 010, 020, 030, etc, are closer together than the 100, 200 set, since the corresponding planes are farther apart.

Let us now choose another set of planes. Figure 2.3 shows the same construction for the (110) set of planes: these planes are even closer together and so the corresponding 110 type points are still farther from the origin. Figure 2.4 repeats this for the (120) planes.

If we continue in this way for all the (*hk*0) planes figure 2.5 is obtained, whence it can be seen that a complete two-dimensional array of points has been generated. This is a section of the *reciprocal lattice*.

Figure 2.6 shows that we can now define a *reciprocal unit cell* with axes **a***, **b*** and interaxial angle γ^*. It can therefore be seen from this that each reciprocal lattice point or *node* corresponds to a *set* of parallel planes. This procedure can obviously be carried out for all possible planes to create a three-dimensional network of points denoted by the indices *hkl* (figure 2.7).

Definition. *Reciprocal lattice vector* g_{hkl} : *The vector $g_{hkl} = h\mathbf{a}^* + k\mathbf{b}^* + l\mathbf{c}^*$ from the origin 000 of the reciprocal lattice to a particular reciprocal lattice point hkl. g_{hkl} is perpendicular to the plane (hkl). The modulus $|g_{hkl}| = 2\pi/d_{hkl}$.*

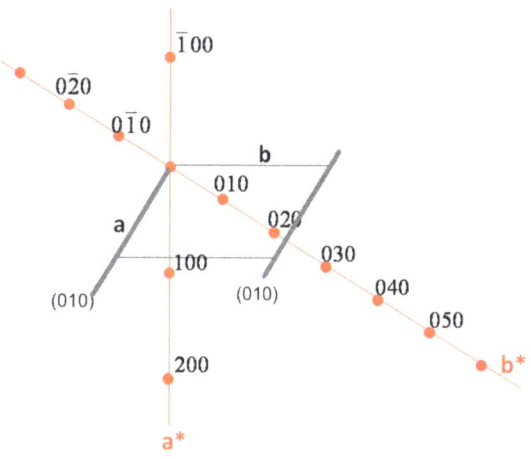

Figure 2.2. Construction of 010, 020, ... reciprocal lattice points.

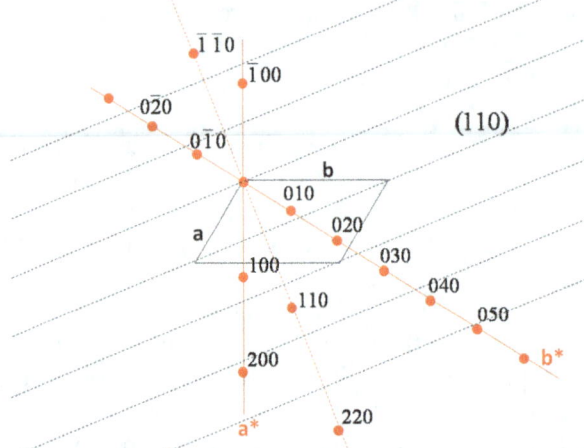

Figure 2.3. Construction of 110, 220 ... reciprocal lattice points. (110) planes marked by dashed lines.

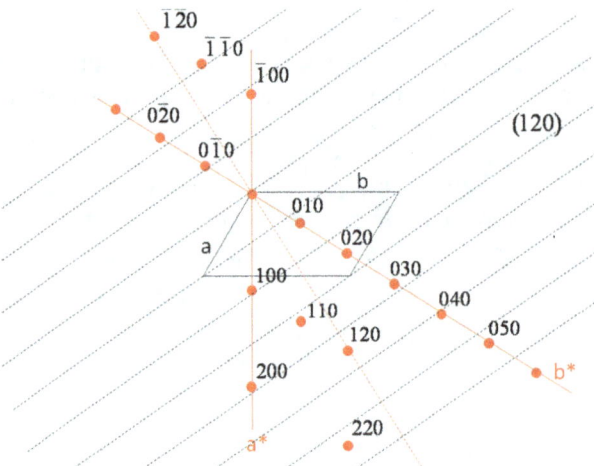

Figure 2.4. Construction of 120, 240 ... reciprocal lattice points. (120) planes marked by dashed lines.

Notation. *It is conventional to denote the indices of reciprocal lattice points by the indices hkl of the relevant planes. Note that no parentheses or brackets are used in specifying reciprocal lattice points.*

Therefore, the reciprocal lattice can be thought of as a three-dimensional lattice of points (or nodes) lying on layers $hk0$, $hk1$, $hk2$, $hk\bar{1}$, $hk\bar{2}$, ...

Conclusion. *For each real lattice array, there corresponds a reciprocal lattice array, whose dimensions are reciprocally related.*

A Journey into Reciprocal Space

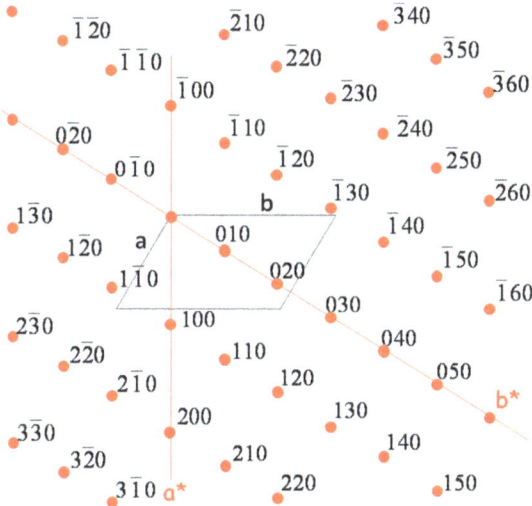

Figure 2.5. A two-dimensional reciprocal lattice.

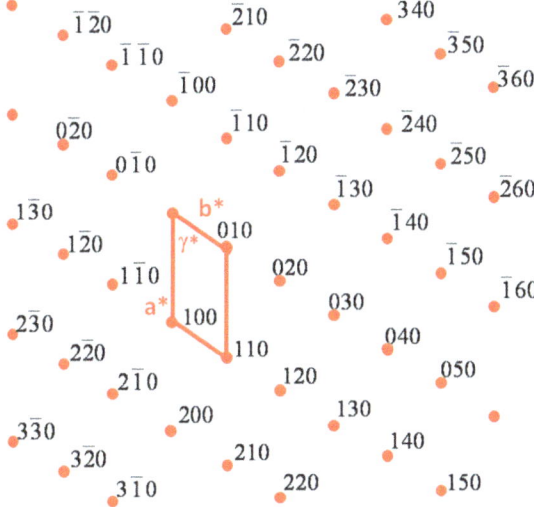

Figure 2.6. Reciprocal unit cell axes and angle.

2.3 Geometrical calculations

Our definitions of the reciprocal axes in terms of vectors in equation (2.3) allow us to calculate a number of useful quantities without the need to perform some sort of elaborate geometrical construction.

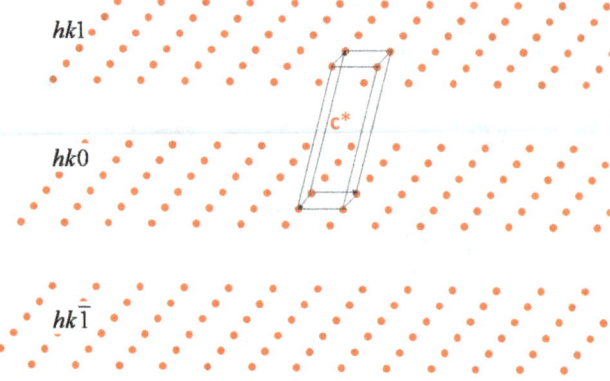

Figure 2.7. Reciprocal lattice in three dimensions: note that one way to visualize this is as a set of parallel layers of reciprocal lattice points.

Metric tensors

It is convenient to express the real and reciprocal axes in matrix form known as the metric tensor. In real space, the metric tensor is given by

$$G = \begin{pmatrix} \mathbf{a}\cdot\mathbf{a} & \mathbf{a}\cdot\mathbf{b} & \mathbf{a}\cdot\mathbf{c} \\ \mathbf{b}\cdot\mathbf{a} & \mathbf{b}\cdot\mathbf{b} & \mathbf{b}\cdot\mathbf{c} \\ \mathbf{c}\cdot\mathbf{a} & \mathbf{c}\cdot\mathbf{b} & \mathbf{c}\cdot\mathbf{c} \end{pmatrix} = \begin{pmatrix} a^2 & ab\cos\gamma & ac\cos\beta \\ ba\cos\gamma & b^2 & bc\cos\alpha \\ ca\cos\beta & cb\cos\alpha & c^2 \end{pmatrix} \quad (2.9)$$

and in reciprocal space correspondingly

$$G^* = \begin{pmatrix} \mathbf{a^*}\cdot\mathbf{a^*} & \mathbf{a^*}\cdot\mathbf{b^*} & \mathbf{a^*}\cdot\mathbf{c^*} \\ \mathbf{b^*}\cdot\mathbf{a^*} & \mathbf{b^*}\cdot\mathbf{b^*} & \mathbf{b^*}\cdot\mathbf{c^*} \\ \mathbf{c^*}\cdot\mathbf{a^*} & \mathbf{c^*}\cdot\mathbf{b^*} & \mathbf{c^*}\cdot\mathbf{c^*} \end{pmatrix} = \begin{pmatrix} a^{*2} & a^*b^*\cos\gamma^* & a^*c^*\cos\beta^* \\ b^*a^*\cos\gamma^* & b^{*2} & b^*c^*\cos\alpha^* \\ c^*a^*\cos\beta^* & c^*b^*\cos\alpha^* & c^{*2} \end{pmatrix} \quad (2.10)$$

The volume V of the real unit cell is given by the determinant of G:

$$V^2 = a^2b^2c^2(1 + 2\cos\alpha\cos\beta\cos\gamma - \cos^2\alpha - \cos^2\beta - \cos^2\gamma) \quad (2.11)$$

Interplanar distances

Suppose we want to work out the interplanar distance d_{hkl} for an (hkl) plane. We simply need to find the modulus of the corresponding reciprocal lattice vector: this can be obtained by taking the scalar product of the vector with itself, thus:

$$|\mathbf{g}_{hkl}|^2 = (g_{hkl})^2 = \left(\frac{2\pi}{d_{hkl}}\right)^2 = (h\mathbf{a^*} + k\mathbf{b^*} + + l\mathbf{c^*}) \cdot (h\mathbf{a^*} + k\mathbf{b^*} + + l\mathbf{c^*}) \quad (2.12)$$

or in metric tensor form

$$(hkl)\begin{pmatrix} a^{*2} & a^*b^* \cos \gamma^* & a^*c^* \cos \beta^* \\ b^*a^* \cos \gamma^* & b^{*2} & b^*c^* \cos \alpha^* \\ c^*a^* \cos \beta^* & c^*b^* \cos \alpha^* & c^{*2} \end{pmatrix}\begin{pmatrix} h \\ k \\ l \end{pmatrix} \qquad (2.13)$$

For the most general case of a triclinic crystal, this is

$$h^2 a^{*2} + k^2 b^{*2} + l^2 c^{*2} + 2kl b^* c^* \cos \alpha^* + 2lh c^* a^* \cos \beta^* + 2hk a^* b^* \cos \gamma^* \qquad (2.14)$$

a rather fearsome result needing calculation of the reciprocal axis values, it must be admitted! However, tedious though it is, it is eminently suitable for calculation by a computer program. For crystal systems with orthogonal axes, the formulae become much easier. Thus, for the orthorhombic system we have

$$a^* = \frac{2\pi}{a} \qquad b^* = \frac{2\pi}{b} \qquad c^* = \frac{2\pi}{c} \qquad (2.15)$$

and then

$$\frac{1}{d_{hkl}^2} = \frac{h^2}{a^2} + \frac{k^2}{b^2} + \frac{l^2}{c^2} \qquad (2.16)$$

For the tetragonal system

$$\frac{1}{d_{hkl}^2} = \frac{h^2 + k^2}{a^2} + \frac{l^2}{c^2} \qquad (2.17)$$

while for the cubic system

$$\frac{1}{d_{hkl}^2} = \frac{h^2 + k^2 + l^2}{a^2} = \frac{N^2}{a^2} \qquad (2.18)$$

where N is an integer.

As a sideline, there is an interesting mathematical aspect of the cubic formula (2.18) in connection with number theory. The sum $h^2 + k^2 + l^2$ cannot equal integers like 7 and 15 (and this accounts for why in a powder x-ray diffraction pattern from a cubic material there are missing lines corresponding to these numbers). In other words, there are certain positive integers that cannot be formed from a sum of three squared integers. How then can one work out which are the other missing positive integers? Fortunately, the formula for determining this was found by Gauss and Legendre, who showed that the sum could not be achieved when

$$N = 4^m(8n + 7) \qquad (2.19)$$

where m and n are positive integers. Table 2.1 can then be constructed for the missing N integers:

Table 2.1. Missing integers from Gauss and Legendre.

m	n	N
0	0	7
0	1	15
0	2	23
1	0	28
0	3	31
0	4	39
0	5	47
0	6	55
1	1	60
0	7	63

Angle between plane normals

To calculate the angle ϕ between any two planes $(h_1k_1l_1)$ and $(h_2k_2l_2)$, or to be more precise, the angle between their normals, just use a scalar product between the relevant reciprocal lattice vectors:

$$(h_1\mathbf{a}^* + k_1\mathbf{b}^* + l_1\mathbf{c}^*) \cdot (h_2\mathbf{a}^* + k_2\mathbf{b}^* + l_2\mathbf{c}^*)$$
$$= |h_1\mathbf{a}^* + k_1\mathbf{b}^* + l_1\mathbf{c}^*||h_2\mathbf{a}^* + k_2\mathbf{b}^* + l_2\mathbf{c}^*|\cos\phi \quad (2.20)$$

Example. *Consider a monoclinic crystal with a = 5.2 Å, b = 8.1 Å, c = 6.4 Å and β = 110°. Calculate the angle ϕ between the normals to the (211) and (22$\bar{3}$) planes.*

Take the scalar product of the reciprocal lattice vectors:
$$(2\mathbf{a}^* + \mathbf{b}^* + \mathbf{c}^*) \cdot (2\mathbf{a}^* + 2\mathbf{b}^* - 3\mathbf{c}^*) = (4a^{*2} + b^{*2} + c^{*2} + 4a^*c^* \cos\beta^*)^{\frac{1}{2}}$$
$$\times (4a^{*2} + 4b^{*2} + 9c^{*2} - 12a^*c^* \cos\beta^*)^{\frac{1}{2}} \cos\phi \quad (2.21)$$

The left-hand side of this equation can then be rewritten as

$$4a^{*2} + 2b^{*2} - 3c^{*2} - 4a^*c^* \cos\beta^* \quad (2.22)$$

and the reciprocal axes and angles can be substituted by the real values using the relationships:

$$a^* = \frac{2\pi}{a \sin\beta}$$
$$b^* = \frac{2\pi}{b}$$
$$c^* = \frac{2\pi}{c \sin\beta} \quad (2.23)$$
$$\beta^* = 180 - \beta$$

Putting in the unit cell values we find that the angle $\phi = 76.55°$.

Angle between plane normal and direction

Again, this is a simple matter using scalar vector products, only here the conditions in equation (2.8) make life easier still. So, let us consider another example using the same monoclinic crystal.

Example. *Calculate the angle ϕ between the normal to the (211) plane and the $[22\bar{3}]$ direction.*

Recall that a direction $[uvw]$ in real space is given by the vector $u\mathbf{a} + v\mathbf{b} + w\mathbf{c}$ and therefore consider

$$(2\mathbf{a}^* + \mathbf{b}^* + \mathbf{c}^*) \cdot (2\mathbf{a} + 2\mathbf{b} - 3\mathbf{c})$$
$$= (4a^{*2} + b^{*2} + c^{*2} + 4a^*c^* \cos \beta^*)^{\frac{1}{2}} (4a^2 + 4b^2 + 9c^2 - 12ac \cos \beta)^{\frac{1}{2}} \cos \phi \quad (2.24)$$

The left-hand side of this equation can then be rewritten, recalling that real and reciprocal cross-products equal zero, according to equation (2.8), as

$$2\pi(4 + 2 - 3) = 6\pi \quad (2.25)$$

Notice, by the way, that the factor of 2π (remember that crystallographers use the value 1 instead) cancels out on both sides of equation (2.24). Putting this together we find that $\phi = 66.43°$.

References

[1] Nespolo M and Souvignier B 2010 The Bravais polar lattice as a didactic tool for diffraction beginners *J. Appl. Crystallogr.* **43** 1144–9
[2] Kittel C 2005 *Introduction to Solid State Physics* 8th edn (Chichester: Wiley)
[3] Ashcroft N W and Mermin N D 1976 *Solid State Physics* vol 2 (Chichester: Wiley)
[4] Simon S H 2013 *The Oxford Solid State Basics* (Oxford: Oxford University Press)

IOP Concise Physics

A Journey into Reciprocal Space
A crystallographer's perspective
A M Glazer

Chapter 3

Diffraction

x-rays ... I am afraid of them. I stopped experimenting with them two years ago, when I came near to losing my eyesight and Dally, my assistant practically lost the use of both of his arms.

Thomas Edison

So far we have learnt how to construct a reciprocal lattice and we have discovered that it enables us to make some complicated geometrical questions easier to attack. However, you will not be surprised to learn that the reciprocal lattice concept has much, much more to offer. We shall now see how we can use it to understand a very practical subject, namely what happens when x-rays, neutrons and electrons are diffracted by a crystalline material.

In 1912, Walter Friedrich, Paul Knipping and Max Laue discovered that x-rays could be diffracted by crystals (figure 3.1), thus proving for the first time that crystal structures consisted of periodic arrays of atoms and molecules with spacings comparable with the wavelength of the x-rays. This also demonstrated that x-rays could be treated as being wave-like in nature (two birds with one stone!). For this discovery, Laue (by this time having been ennobled as von Laue!) received the 1914 Nobel prize in Physics.

3.1 Laue equations

In order to explain the observed x-ray photographs, Laue treated the diffraction effect in a very simple, but correct, way. Consider x-rays striking a row of atoms spaced a distance **a** apart and being scattered in a different direction. Figure 3.2 shows this together with a vector diagram with the unit vector $\mathbf{s}_0$ for the incident beams and $\mathbf{s}$ for the scattered beams. It is obvious that there is now a path difference between the beams separated by the distance **a** and this is given simply by

$$\mathbf{a} \cdot \mathbf{s} - \mathbf{a} \cdot \mathbf{s}_0 = \mathbf{a} \cdot (\mathbf{s} - \mathbf{s}_0) \tag{3.1}$$

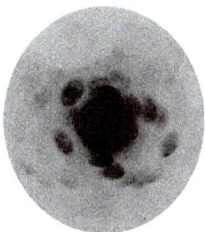

Figure 3.1. Friedrich and Knipping's first successful photograph showing that x-rays can be diffracted by a crystal of copper sulphate pentahydrate.

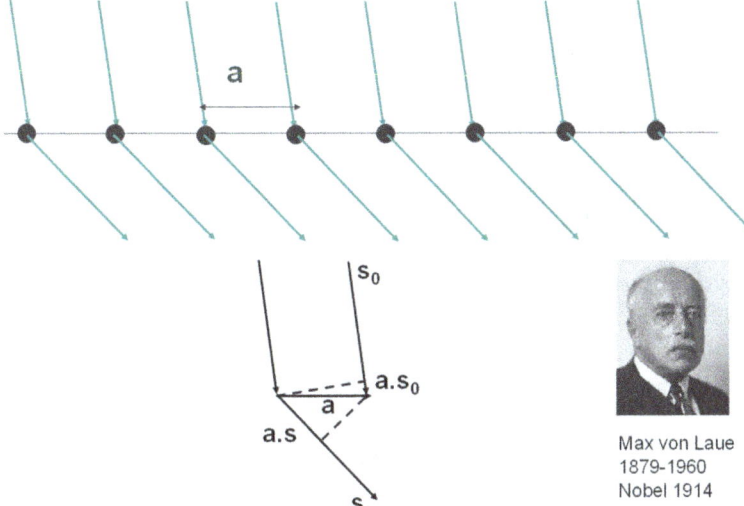

Max von Laue
1879-1960
Nobel 1914

Figure 3.2. Derivation of the Laue equations.

For constructive interference, this must be equal to an integer h of wavelengths, and so

$$\mathbf{a} \cdot (\mathbf{s} - \mathbf{s}_0) = h\lambda \tag{3.2}$$

Therefore, in three dimensions we can write the following set of equations

$$\begin{aligned}\mathbf{a} \cdot (\mathbf{s} - \mathbf{s}_0) &= h\lambda \\ \mathbf{b} \cdot (\mathbf{s} - \mathbf{s}_0) &= k\lambda \\ \mathbf{c} \cdot (\mathbf{s} - \mathbf{s}_0) &= l\lambda\end{aligned} \tag{3.3}$$

Definition: *The equation (3.3) are known as the **Laue equations**.*

3.2 Bragg's Law

We now turn to one of the most famous equations in physics: Bragg's Law. The Law was derived by William Lawrence Bragg in 1912, at the young age of 22, as a simple way in which to work out the angle of diffraction from any set of crystal planes, and can be

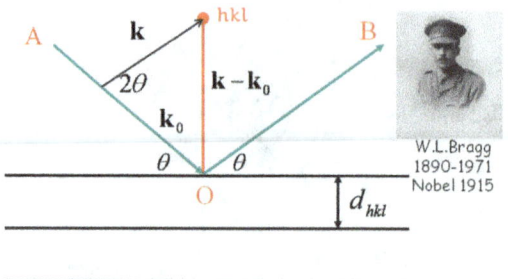

Figure 3.3. Derivation of Bragg's Law.

understood by reference to figure 3.3. Be aware that the derivation given here is not quite the conventional proof that you will see in most crystallography text-books, but it is done deliberately this way to emphasize the use of the reciprocal lattice concept. Consider a set of planes in the crystal with interplanar spacing d_{hkl}. Now, Bragg's brilliant idea was to treat the diffraction process as one of reflection of x-rays from crystal planes, as if the planes acted like mirrors. This, of course, is not strictly true, as diffraction and reflection are different processes, but nonetheless, it does lead to a useful formula that at least produces the correct *positions* of the peaks in the diffraction pattern.

Notation: *Because Bragg treated diffraction in terms of reflection, it is internationally accepted that the diffraction spots arising from scattering by crystal planes are called* **reflections**.

Let a beam AO of radiation strike a crystal plane at an angle θ (figure 3.3). This angle is conventionally known as the Bragg angle. The beam is then considered to be 'reflected' through an equivalent angle along OB just like light reflected by a mirror.

Now we draw a perpendicular to the planes and mark the corresponding reciprocal lattice point *hkl* at a distance $2\pi/d_{hkl}$ (recall our construction of the reciprocal lattice by drawing perpendiculars to planes). Now draw a line marked **k** parallel to the reflected beam OB to meet this reciprocal lattice point, and then we see that this makes an angle 2θ to the incident beam AO. The result is a triangle of wave-vectors with incident wave-vector $\boldsymbol{k}_0$ and outgoing wave-vector $\boldsymbol{k}$. The reciprocal lattice vector $\boldsymbol{g}_{hkl}$ (magnitude equals $2\pi/d_{hkl}$) is then given by

$$\boldsymbol{k} - \boldsymbol{k}_0 = \boldsymbol{g}_{hkl} \tag{3.4}$$

Definition: *The vector* $\boldsymbol{k} - \boldsymbol{k}_0$ *is called the* **scattering vector**.

Therefore

$$(\boldsymbol{k} - \boldsymbol{k}_0) \cdot (\boldsymbol{k} - \boldsymbol{k}_0) = \boldsymbol{g}_{hkl} \cdot \boldsymbol{g}_{hkl} \tag{3.5}$$

and this leads to

$$k^2 + k_0^2 - 2kk_0 \cos 2\theta = \frac{4\pi^2}{d_{hkl}^2} \tag{3.6}$$

But for elastic scattering, the wavelength λ is unchanged, and so the moduli of the wave-vectors are given by

$$k = k_0 = \frac{2\pi}{\lambda} \tag{3.7}$$

Rearranging then gives us the well-known Bragg equation

$$\lambda = 2d \sin \theta \tag{3.8}$$

Those of you who have met the Bragg equation before will probably have seen it in the following form

$$n\lambda = 2d \sin \theta \tag{3.9}$$

where n is a positive integer and represents the *order* of diffraction. Now, it is my experience that removing the factor n as in equation (3.8) seems to confuse students when they meet it the first time and so it is worth explaining the reasoning behind this. As you will no doubt know, when light is diffracted from, say, a slit, one obtains a central maximum of intensity with weaker periodic sideband peaks. It is then convenient to give each intensity peak a serial number denoting the order of diffraction.

Suppose we consider x-rays being diffracted from, say, the (130) set of parallel planes. One way of thinking about this is to say that we obtain a set of diffracted beams at increasing angles corresponding to the increasing values of n. But another way (figure 3.4) is to forget about the order n and instead think about reflection

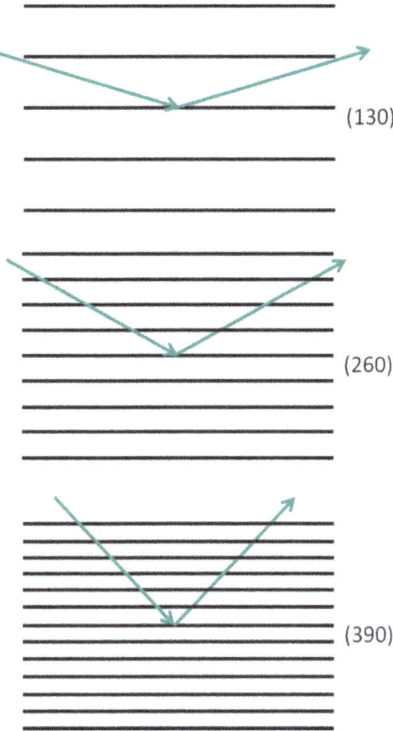

Figure 3.4. Diffraction from the (130) planes.

(diffraction) from planes (130), (260), (390)..... In other words, we consider hypothetical planes half-way, third-way etc with respect to (130) corresponding to spacings d_{130}, d_{260}, d_{390}....

So why should we prefer this to the more usual use of talking about an order as in optical diffraction? The point is that a crystal is a three-dimensional object described by a three-dimensional lattice. Therefore, if we think of sets of different planes such as (100), (010), (130) and so on, we could write the following:

	Orders				
(100)	1	2	3	4	5
(010)	1	2	3	4	5
(130)	1	2	3	4	5

in which case, we would have to talk about first, second, third etc orders for (100) planes and a different set of first, second, third etc orders for (010), and still different again for (130).

This is clearly a messy way of labelling the reflections. Instead, if we invent hypothetical planes with corresponding interplanar spacings we can then label the reflections like this

100	200	300	400	500
010	020	030	040	050
130	260	390	4 12 0	5 15 0

and these simply correspond to the labels for the reciprocal lattice points we constructed in the previous chapter! (Remember that we do not use parentheses when labelling the reciprocal lattice points). You can now begin to appreciate that the reciprocal lattice construction is telling us something about the diffraction pattern obtained from a single crystal.

Definition: *The equation $\lambda = 2d \sin \theta$ is known as* **Bragg's Law**.

Note: *The angle through which the incident beam is diffracted is equal to 2θ, i.e., twice the Bragg angle.*

Returning for the moment to the Laue equations, suppose we take the scalar product of the unit cell vector **a** and the reciprocal lattice vector $\mathbf{g}_{hkl}$, thus

$$\mathbf{a} \cdot \mathbf{g}_{hkl} = \mathbf{a} \cdot (h\mathbf{a}^* + k\mathbf{b}^* + l\mathbf{c}^*) = 2\pi h \tag{3.10}$$

Comparing this with the Laue equations (3.2) and (3.4), we note that $\mathbf{s} - \mathbf{s}_0$ is parallel to the reciprocal lattice vector $\mathbf{g}_{hkl}$. Thus, the Laue equation is consistent with the Bragg idea of treating the crystal diffraction as if it were reflection from planes—equation (3.10) and so

$$\frac{2\pi}{\lambda}(\mathbf{s} - \mathbf{s}_0) = \mathbf{g}_{hkl} \tag{3.11}$$

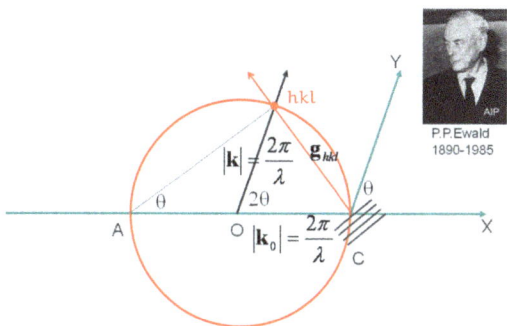

Figure 3.5. The Ewald sphere construction.

3.3 The Ewald sphere

In 1912 when Paul Ewald introduced the reciprocal lattice he showed how this could be used to visualize the formation of a diffraction pattern from a crystal. In fact, we have already seen this in the way Bragg's Law was derived in section 3.2 by making use of wave-vectors. This can be developed still farther by use of a marvelous construction due to Ewald. Take a look at figure 3.5.

First of all, consider a beam of radiation AX reflected (diffracted) along the direction CY by a set of parallel planes C in a crystal. Now, construct a sphere of radius, whose radius is inversely related to the wavelength of the x-rays, such that it is centred at O with the crystal C on the circumference. This is the so-called **Ewald sphere** or **sphere of reflection**. Perpendicular to the set of planes mark in a line, which, by definition, is in the direction of the reciprocal lattice vector g_{hkl} for the planes. Then draw in a line from O parallel to the outgoing ray CY. This makes an angle of 2θ to the incident beam and meets the circumference of the sphere at the same point as the reciprocal lattice vector. Suppose now that this intersection point happens to coincide with the *hkl* reciprocal lattice point for the planes. Draw in the line from A to the reciprocal lattice point and the result is a right-angled triangle. Hence, we get

$$\sin\theta = \frac{g_{hkl}}{4\pi/\lambda} = \frac{\lambda}{2d_{hkl}} \qquad (3.12)$$

which is our old friend Bragg's Law! So, what does this mean exactly?

Well, what it says is that, if a reciprocal lattice point happens to lie on the surface of the Ewald sphere, Bragg's Law is obeyed and there will be a diffracted beam from the crystal at an angle 2θ in the direction parallel to the line from O to the reciprocal lattice point. Conversely, if the reciprocal lattice point does not lie on the sphere then the corresponding set of planes will not be in the correct orientation for diffraction to occur.

3.4 Lost in reciprocal space?

The Ewald sphere construction is going to give us a very convenient means to enable us to find our way around reciprocal space and determine what we shall see in a diffraction pattern. I used this myself many years ago, for instance, when I wanted to determine the presence of an extremely weak reflection. To do this it was necessary to oscillate the crystal through just 1° for about one week. This is a tall order but we managed it by carefully drawing a reciprocal lattice to scale and using the Ewald sphere to find out where to set the crystal (this was before modern automatic diffractometers made all this trivial!).

Stationary crystal

Let us see how we can use the Ewald construction to understand the diffraction pattern that can be obtained from a crystal. Suppose that we use a fixed single wavelength for the incident radiation and allow it to fall on a stationary crystal (figure 3.6). In this diagram the $hk0$ section of a reciprocal lattice is shown with an x-ray (or neutron or electron) beam coming from the left (coloured green of course!).

It can be seen that in this section only reciprocal lattice points, $0\bar{3}0$ and $2\bar{3}0$ actually lie on the Ewald sphere (I have coloured them green to emphasize this). This results in two x-ray beams diffracted through the angles $2\theta_{0\bar{3}0}$ to the left of the incident beam and $2\theta_{2\bar{3}0}$ to the right. So, if we use a detector (film or counter) to observe x-ray reflections within this reciprocal lattice plane we would only observe these two reflections as spots or peaks in intensity: none of the other reflections would be observable in this reciprocal lattice layer.

In figure 3.7 we see what happens in three dimensions. The drawing shows reciprocal lattice layers perpendicular to the c axis. We can determine the indices of the layers in the following way. Recall that the c-axis direction is in fact equivalent to [001], and so consider the scalar product

$$\mathbf{c} \cdot (h\mathbf{a}^* + k\mathbf{b}^* + l\mathbf{c}^*) = 0 \tag{3.13}$$

for layers perpendicular to [001]. This gives the result

$$2\pi l = 0 \tag{3.14}$$

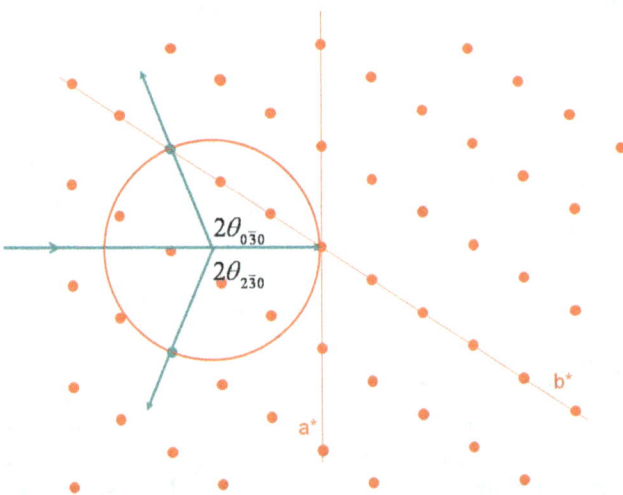

Figure 3.6. $hk0$ section of reciprocal lattice and Ewald Sphere for stationary crystal.

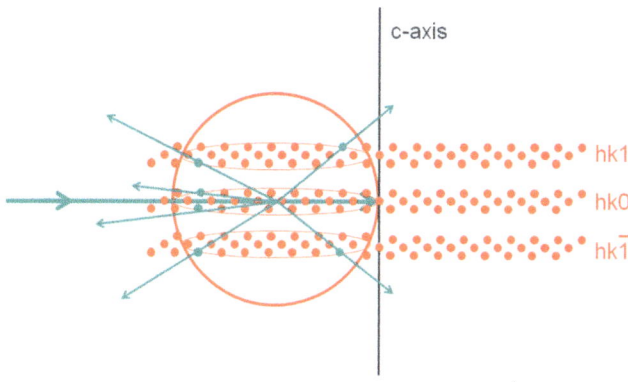

Figure 3.7. Ewald sphere for stationary crystal.

and therefore, the layers have the indices $hk0$, $hk1$, $hk\bar{1}$, $hk2$, $hk\bar{2}$, $hk3$, $hk\bar{3}$, ...

It can be seen that the layers cut through the Ewald sphere, but still only very few reciprocal lattice points actually are on the surface of the sphere. This shows that with a stationary crystal like this and a monochromatic source of incident radiation, in general, very few reflections would be expected to be observed.

Now this points to a lesson about the original experiment carried out back then in 1912 by Friedrich and Knipping. They used a stationary crystal, and yet observed a sufficient deflection of the x-ray beam to realize that they had found evidence of diffraction. They soon were able to obtain beautiful photographs from ZnS showing many spots (reflections) arranged in patterns showing underlying symmetry. The problem that they then faced was: how to interpret what they had seen? The first crystal that they used for their experiment was $CuSO_4.5H_2O$, because Laue mistakenly thought that any diffraction would be the result of scattering by a single wavelength due to secondary x-rays from the copper atoms. Had this really been the case our analysis shows that they probably would not have obtained the first diffraction pattern and they would have returned home disappointed[1]. So, luck was with Friedrich and Knipping on the occasion of their seminal experiment and handed Laue his Nobel prize. This illustrates how great scientific advances so often occur by pure chance.

Both father and son, William Henry and William Lawrence Bragg, soon set about trying to understand the 'curious' photographs obtained by Laue (figure 3.8). Having realized that, if they assumed that the x-rays were polychromatic, they could use this knowledge to work out the first atomic structures of crystals such as NaCl. W L Bragg's model of alternating Na and Cl atoms stretching out in all directions was not without controversy however: see the letter to *Nature* in 1927 from Professor H E Armstrong [1], below. Many chemists thought that NaCl should

---

[1] Laue tried but failed to interpret the observed photographs fully, still persisting in the belief that only one or possibly five wavelengths were involved, but it was W L Bragg who showed that the diffraction patterns could be explained if the incident radiation consisted of a continuum of wavelengths. Surprisingly, Laue should have realized his mistake, since Friedrich and Knipping obtained a good diffraction pattern from a crystal of diamond, where secondary x-rays would not be expected to play a role.

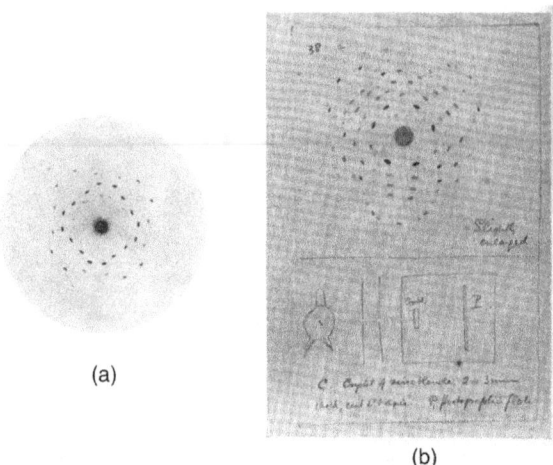

Figure 3.8. (a) One of Friedrich and Knipping's original x-ray photographs of ZnS. (b) Facsimile of one of W H Bragg's diagrams of Laue's 'curious' x-ray effect (1912).

consist of molecules and for a long time resented the fact that mere physicists were telling them that they were wrong! Never mind—both father and son Bragg shared the 1915 Nobel prize for their discovery and the structure of NaCl has stood the test of time!

Poor Common Salt

'Some books are lies frae end to end' says Burns. Scientific (save the mark) speculation would seem to be on the way to this state!.....Professor W.L. Bragg asserts that 'In Sodium Chloride there appear to be no molecules represented by NaCl. The equality in number of sodium and chlorine atoms is arrived at by a chess-board pattern of these atoms; it is a result of geometry and not of a pairing of the atoms'.

This statement is more than 'repugnant to common sense'. It is absurd to the n...th degree, not chemical cricket. Chemistry is neither chess nor geometry, whatever X-ray physics might be. Such unjustified aspersion of the molecular character of our most necessary condiment must not be allowed any longer to pass unchallenged. It were time that chemists took charge of chemistry once more and protected neophytes against the worship of false gods; at least taught them to ask for something more than chess-board evidence.

Professor H.E. Armstrong, Letter to Nature 1927

Oscillating and rotating crystal

We have seen that with a stationary crystal and a monochromatic beam the chances of diffraction are fairly minimal. In order to capture some more diffraction maxima, we shall need to move the crystal in some way.

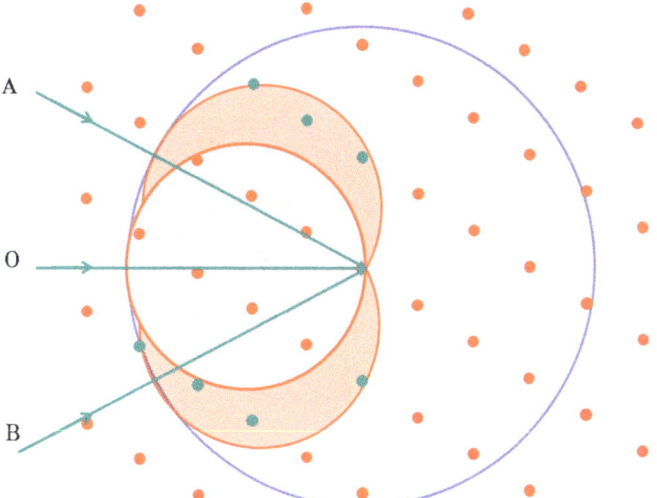

Figure 3.9. Effect of oscillating a crystal through an angle.

Referring to figure 3.9 imagine now that we oscillate the crystal through an angle about the axis perpendicular to the reciprocal lattice layer shown. For convenience, this is shown here by keeping the reciprocal lattice fixed and making the incident beam oscillate between the positions A and B, thus causing the Ewald sphere to oscillate too. It can now be seen that the shaded areas ('lunes') include several reciprocal lattice points (green), and this means that during the course of the oscillation, corresponding reflections will occur on a suitably placed detector. It should be obvious, that, in general, in an oscillation experiment, different reflections appear on either side of the x-ray beam so that the observed pattern will normally be asymmetric, unless one happens to oscillate either side of a prominent symmetry axis direction.

If one has a crystal with a very large unit cell, such as is typically found in protein crystals, even a small oscillation range will give rise to a large number of observed reflections (figure 3.10). In this example, the density of reflections is such that you can see the reflections lying on circles about the origin of the photograph corresponding to circular cuts through the Ewald sphere. Therefore, one way of thinking about this is to realize that the diffraction pattern is actually an imprint of the reciprocal lattice, distorted by the geometry of the experimental arrangement. The varying intensities of the spots arise from the different atomic arrangements on the planes giving rise to the reflections, and it is this that eventually enables crystallographers to solve the crystal structure.

It can also be seen from figure 3.9 that if we rotate the crystal (or beam) through a complete 360° all reciprocal lattice points within a large circle (marked in blue) will at some time pass through the circumference of the Ewald sphere and thus create a reflection on the detector. In three dimensions, this limiting circle is a sphere.

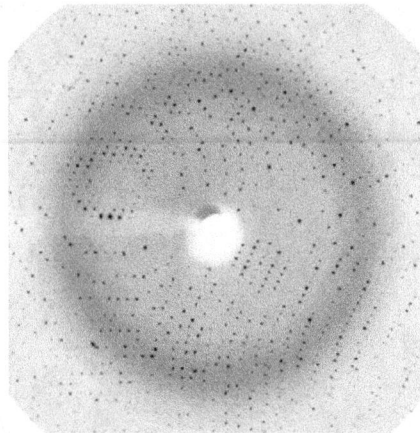

Figure 3.10. x-ray photograph obtained using a synchrotron source from a protein crystal oscillating over a very small angle. Reproduced with permission of the International Union of Crystallography.

Definition: *The limiting sphere is one that contains all the possible reciprocal lattice points for a particular wavelength, where*

$$d_{hkl} \geq \frac{\lambda}{2} \qquad (3.15)$$

Note that in a rotation experiment like this the same reflections are captured on each side of the incident beam and so the observed diffraction pattern will appear to be symmetric about the axis of rotation. One of the disadvantages of the oscillation/rotation method is that the diffraction spots will often overlap, thus making it difficult to measure each independent reflection. Before the modern era of computerized diffractometers, when I started crystallographic research, diffraction patterns were recorded on photographic film, and so in order to remove the overlap problem, moving film methods were used to spread the reflections out. The two most common were the so-called precession and Weissenberg techniques.

Polycrystalline powder

Suppose that, instead of using a single crystal, radiation is diffracted by a polycrystalline or powdered specimen. Now this is simply a collection of small crystallites, randomly oriented in an ideal case. In reciprocal space, this can be thought of by effectively spinning the reciprocal lattice about its origin. Figure 3.11 shows that by rotating the reciprocal lattice about its centre each reciprocal lattice point traces out a circle (coloured grey), so that a flat photographic film place perpendicular to the incident beam will see circular rings of diffracted intensity projected onto it.

Clearly if the powder does not consist of randomly oriented crystallites the intensity around these rings will be uneven. This type of effect is known as **preferred orientation**. If the powder only consists of a very small number of crystallites, then the rings will appear to be spotty. Thus, when carrying out a powder diffraction

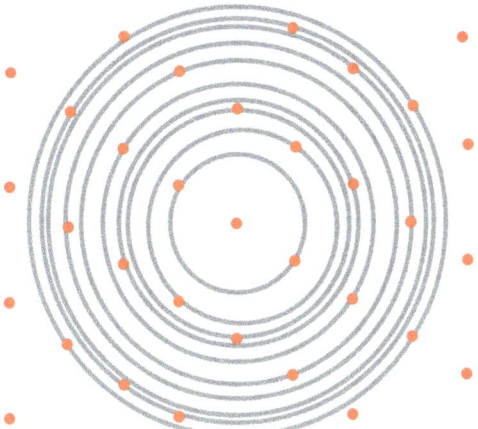

Figure 3.11. Effect of rotating the reciprocal lattice about its origin. Here the incident beam is assumed to be perpendicular to the page.

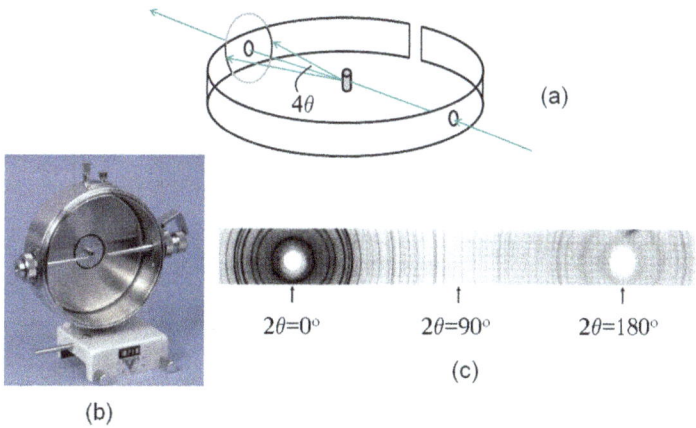

Figure 3.12. The Straumanis method of powder diffraction (a) Schematic diagram. X-ray film is placed around the inside of a cylindrical camera. In this method two holes are punched in the film for the incoming and outgoing incident beams. (b) Commercial powder camera. (c) Film exposed to x-rays. The two holes are centred at $2\theta = 0°$ and $180°$. By measuring rings either side of the holes it is possible to determine the true positions of $2\theta = 0°$ and $180°$ and then use this to correct for errors in the film radius caused by shrinkage during the photographic development process.

measurement it is usual to grind the solid up into fine particles of about 1–10 μm size. These days, rather than use photographic film (figure 3.12) to observe the powder rings, an automatic diffractometer (figure 3.13) is normally used in which a counter is scanned from low to high 2θ within a single plane. This then cuts through the rings to produce a one-dimensional pattern of intensities. A typical example of a powder pattern collected on a modern powder diffractometer is shown in figure 3.14.

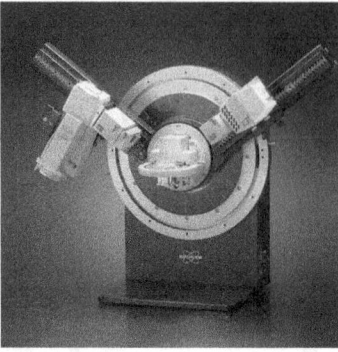

Figure 3.13. Example of an x-ray powder diffractometer manufactured by Bruker AXS.

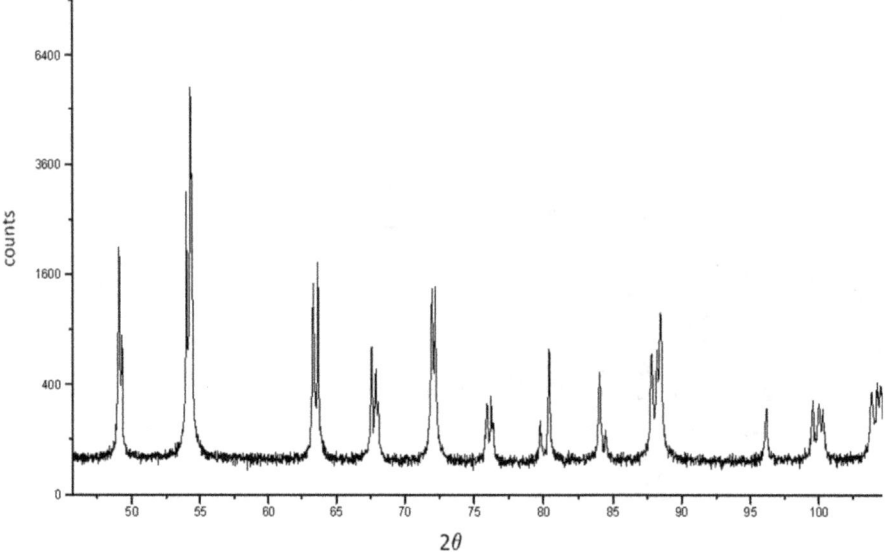

Figure 3.14. Part of an x-ray powder pattern of lead zirconate-titanate (collected with a Panalytical diffractometer and X'Celerator detector).

Laue diffraction

If we go back to the original experiment of Friedrich and Knipping, we can now understand why they were able to obtain their first x-ray diffraction photographs using a stationary crystal. In figure 3.15 we assume that the incident radiation is polychromatic with wavelengths ranging from small to large, thus creating a continuum of Ewald spheres with radii from large to small. Within this range reciprocal lattice points will cut through the surfaces of the Ewald spheres to give rise to diffraction spots. The result on a photographic film is many diffraction spots, each spot arising from a particular wavelength.

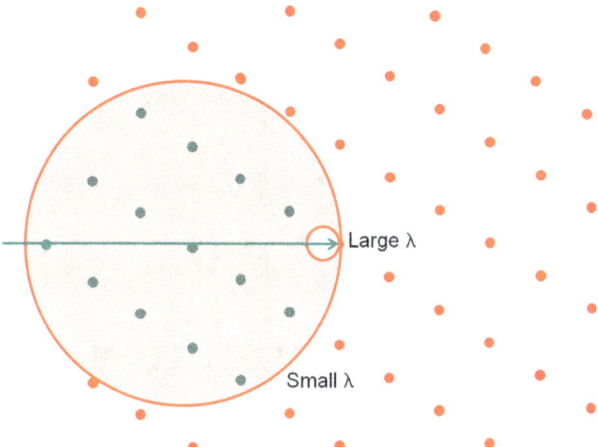

Figure 3.15. The Ewald sphere and Laue diffraction.

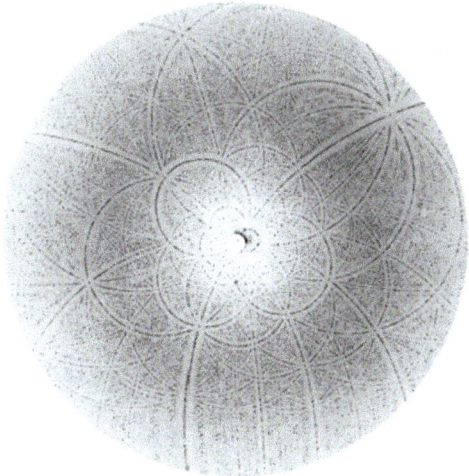

Figure 3.16. Example of a Laue photograph taken using a synchrotron for a crystal of the protein Rubisco [3]. Reproduced with permission of the International Union of Crystallography.

Photographs of this sort have traditionally been used for orienting crystals and establishing crystal quality, either in transmission or in back-reflection geometry. My PhD supervisor (Kathleen Lonsdale) always insisted on seeing the Laue photographs before all else, and I was in trouble if I did not have them to hand!

Laue photography is particularly suited to x-ray synchrotron radiation (figure 3.16) since the source of radiation is completely polychromatic and intense, enabling diffraction information to be collected in microseconds from minute crystals, and has even been used to solve crystal structures, since so many reflections are collected at once. A good review has been given by Ren *et al* [2]

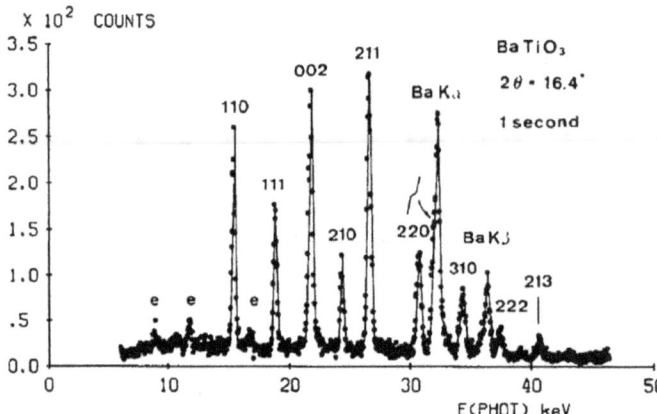

Figure 3.17. Diffraction spectrum of BaTiO$_3$ (3.7 GeV, 14 mA, $2\theta = 16.4°$) from synchrotron source [4]. Letters e denote escape peaks. Note inclusion of Ba Kα_1 fluorescence line. Reproduced with permission of the International Union of Crystallography.

Energy-dispersive diffraction

One of the ways in which polychromatic sources have been used to obtain diffraction information is by collecting the diffracted intensity with a solid-state detector capable of carrying out an energy analysis of the diffracted x-rays. An example of this is in the use of synchrotron radiation (figure 3.17) on a powder sample with the detector set at a fixed 2θ angle. The detector is connected to a multi-channel analyzer which displays the powder pattern as numbers of counts against energy E in keV. Peaks then appear at different values of energy to produce a so-called energy-dispersive powder pattern. Thus, one effectively turns Bragg's Law back to front.

$$E = \frac{hc}{\lambda} = \frac{hc}{2\sin\theta} \cdot \frac{1}{d_{hkl}} \propto \frac{1}{d_{hkl}} \qquad (3.16)$$

where h is Planck's constant and c is the speed of light. The real advantage of this technique is that the sample and detector are stationary making it easier to access a sample housed in some environmental cell for which there is a limited view. This has proved particularly useful in the study of high-pressure phases of materials, where the powder sample is placed in a fluid inside holes in a metal gasket between two opposing diamond crystals that exert high pressure. However, the main problem with using a solid-state detector in this way is that the energy resolution is generally poor and so it cannot compete with ordinary monochromatic angle-scanning methods.

There is another way to carry out energy-dispersive diffraction by scanning through different wavelengths with a crystal monochromator and a conventional detector at a fixed angle (figure 3.18). In this case, the almost perfectly parallel white beam of a synchrotron is used together with a silicon crystal monochromator cut to form two parallel leaves. Rotation of this channel-cut crystal changes the wavelength of the x-rays passed and at the same time keeps the outgoing monochromatic beam parallel to the incident beam, with only a slight change in height. This, in

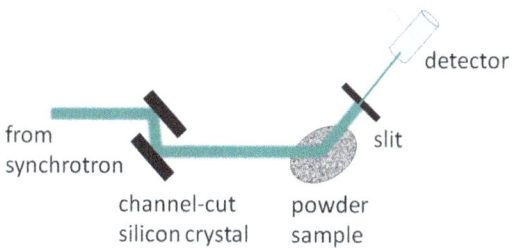

Figure 3.18. High-resolution energy-dispersive diffraction with synchrotron radiation and a channel-cut Si crystal.

principle, provides a resolution of about 1 part in 10^4 given by the scanning monochromator rather than by the detector.

As with x-rays, in neutron diffraction there are basically two ways in which data can be collected. First of all, in the normal angle-dispersive method, neutrons, obtained typically from a nuclear reactor, are diffracted by a monochromator crystal to produce a single wavelength and then shone onto the sample. A detector is then scanned through 2θ. The second way is again an energy-dispersive method, where a pulsed 'white' neutron source is used from a neutron spallation source: here a polychromatic neutron beam is incident on a crystal and a bank of detectors is used to determine the time-of-flight of the scattered neutrons at various fixed angles. The momentum of a neutron with velocity v_N is given by

$$p = m_N v_N = m_N \left(\frac{l}{t}\right) = \frac{h}{\lambda} \qquad (3.17)$$

where l is the distance traversed by the scattered neutron and t its time-of-flight. Then

$$d_{hkl} = \frac{\lambda}{2 \sin \theta} = \frac{h}{2p \sin \theta} = \frac{ht}{2m_N l \sin \theta} \propto t \qquad (3.18)$$

Thus, a detector set at a fixed angle measuring the number of neutrons arriving for different times t produces a diffraction pattern in which d_{hkl} is proportional to time-of-flight. Figure 3.19 shows a typical powder pattern collected in this way. Compare this pattern with that shown in figure 3.14 which shows the same reflections (but plotted left to right). One of the advantages of this type of data collection is that it can rapidly access extremely high hkl values with very high resolution.

3.5 Intensity

In order to understand diffraction of radiation by a crystal we start by considering the more familiar problem of how an optical lens creates an image of an object. Figure 3.20 is a simplified ray diagram showing how light rays pass through a glass lens. Start by following the rays from the top and bottom of the Eiffel Tower at the left. The lens causes all these rays to pass through the focal plane of the lens to form an inverted image at the right (the image plane). However, notice that rays from the top, bottom, and hence everywhere between, coming from the object cross at a point

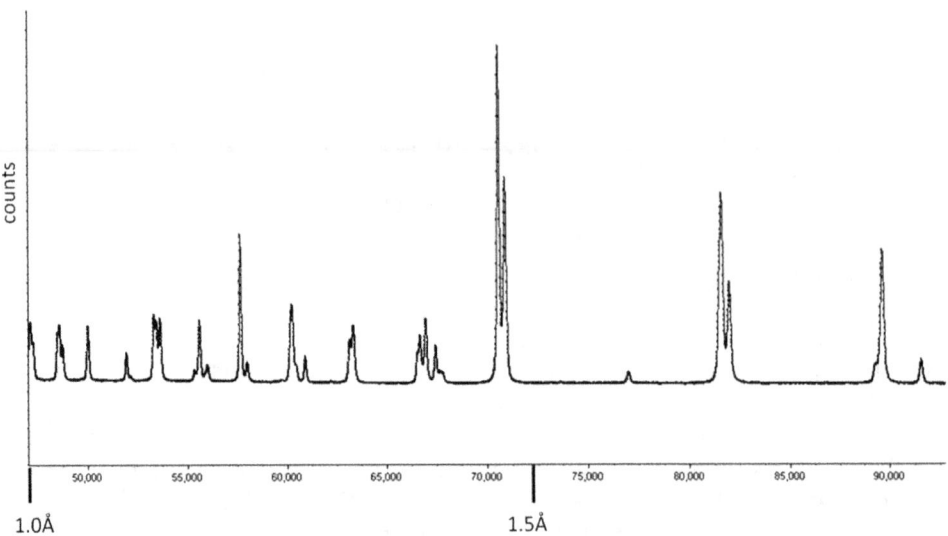

Figure 3.19. Part of time-of-flight powder pattern of lead zirconate titanate collected at the ISIS spallation source at the Rutherford-Appleton Laboratory (horizontal axis marked in channel numbers).

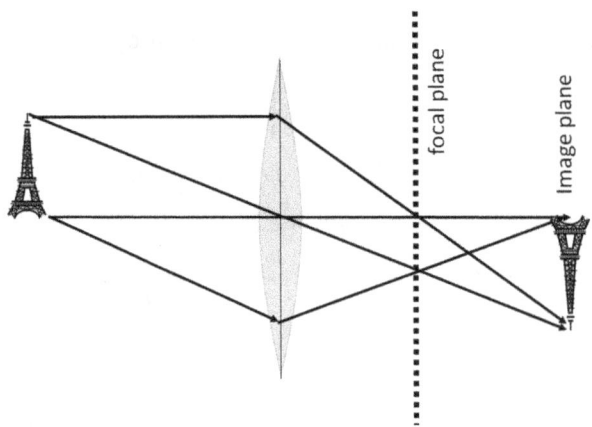

Figure 3.20. Focussing of light by a lens.

on the focal plane. Thus, every point on the focal plane is in receipt of information from the *whole* of the object at the same time.

You can see this for yourself by a simple experiment using a slide projector. First set up a projector showing a focused image of a slide on a screen (figure 3.21). Now remove the focusing lens and go up to the screen and place the lens anywhere in the lit area of the screen. The lit area contains everywhere all the amplitudes and phases of light waves scattered by the image on the slide. You will then observe a small image of the whole of the slide, and as you move the lens around the image will follow, showing that when the lens is removed the light arriving at the screen contains all the information about the slide at every place within the lit area. This

Figure 3.21. An image can be formed from anywhere within the projector beam.

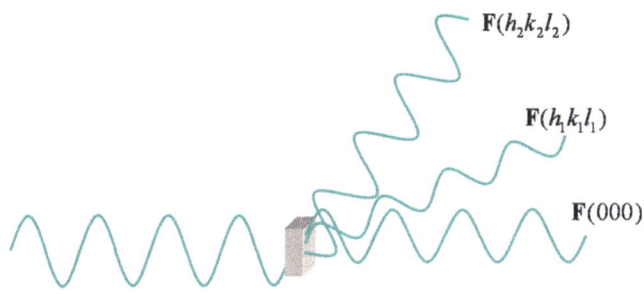

Figure 3.22. Amplitudes and phases of diffracted waves.

demonstrates nicely that the purpose of the lens is to combine all the amplitudes and phases of the light waves from the object to form the image.

Suppose now we consider what happens when x-rays or neutrons are diffracted by an object. In this case, to all intents and purposes, there are no lenses available (in electron diffraction magnetic lenses are available in an electron microscope and so in this case direct images can be formed), and therefore radiation will be scattered into all directions and cannot directly form an image. When the radiation is incident on a crystal waves are diffracted (figure 3.22) by the (*hkl*) planes at angles given by Bragg's Law and will have amplitudes **F**(*hkl*) and phases ϕ(*hkl*). In order to calculate the intensity of an *hkl* reflection for any crystal structure, it is first necessary to find the formula for the amplitude and phase of the scattered wave at the *hkl* reciprocal lattice nodes. The intensity then at the point *hkl* is given by:

$$I(hkl) \propto \mathbf{F}(hkl)\mathbf{F}(hkl)^* = |\mathbf{F}(hkl)|^2 \qquad (3.19)$$

In crystal structure determination one attempts to reconstruct the scattering object (the atoms making up the crystal structure) from the observed diffraction pattern: to do this one needs the amplitudes and the phases for each hkl reflection. However, in x-ray and neutron diffraction, it is normally only the intensity that one observes in the detector, because no lens is available, and so, as can be seen from equation (3.19), the phase information is lost. This difficulty in determining the appropriate phases is known as the *phase problem* to crystallographers and constitutes the main difficulty in determining crystal structures. We shall see later some of the inventive ways in which this has been overcome.

3.6 Fourier transformation

In order to be able to calculate diffraction amplitudes, and hence intensities, we first have to describe the mathematics that will be needed. Calculation of the amplitude of scattering is best done through the use of the Fourier transform. This is defined by the following equation:

$$A(\mathbf{Q}) = \int_{\text{direct}} \rho(\mathbf{r}) e^{i\mathbf{r} \cdot \mathbf{Q}} \, d\mathbf{r}^3 \tag{3.20}$$

where $\rho(\mathbf{r})$ is a function that typically represents a density of some sort at distances given by $\mathbf{r}$ (for example in the case of x-ray scattering this will be the electron density in a crystal). The vector $\mathbf{r}$ therefore is a vector in direct space. Also

$$|\mathbf{Q}| = \frac{2\pi}{\lambda} |\mathbf{s} - \mathbf{s}_0| = \frac{4\pi \sin \theta}{\lambda} \tag{3.21}$$

which is a vector in what we may call 'Fourier transform space', i.e., yet another name for reciprocal space. The integral in equation (3.20) is over all direct space (in the language of crystals this would be the real space of the crystal structure). Now let us consider the Fourier transform of $A(\mathbf{Q})$:

$$G(\mathbf{r}') = \int_{\text{Fourier}} A(\mathbf{s}) e^{-i\mathbf{r}' \cdot \mathbf{Q}} \, d\mathbf{Q}^3 \tag{3.22}$$

The minus sign in the exponential term is there because now we are transforming from Fourier space back to direct space. Therefore

$$\begin{aligned}
G(\mathbf{r}') &= \int_{\text{Fourier}} \left[\int_{\text{direct}} \rho(\mathbf{r}') e^{i\mathbf{r} \cdot \mathbf{Q}} \, d\mathbf{r}'^3 \right] e^{-i\mathbf{r}' \cdot \mathbf{Q}} \, d\mathbf{Q}^3 \\
&= \int_{\text{Fourier}} \int_{\text{direct}} \rho(\mathbf{r}') e^{i(\mathbf{r}-\mathbf{r}') \cdot \mathbf{Q}} \, d\mathbf{r}'^3 \, d\mathbf{Q}^3 \\
&= \int_{\text{direct}} \rho(\mathbf{r}') \, d\mathbf{r}'^3 \int_{\text{Fourier}} e^{i(\mathbf{r}-\mathbf{r}') \cdot \mathbf{Q}} \, d\mathbf{Q}^3
\end{aligned} \tag{3.23}$$

Now it can be shown that

$$\delta(\mathbf{r} - \mathbf{r}') = \int_{\text{Fourier}} e^{i(\mathbf{r}-\mathbf{r}') \cdot \mathbf{Q}} \, d\mathbf{Q}^3 \tag{3.24}$$

where $\delta(\mathbf{r} - \mathbf{r}')$ is the three-dimensional Dirac delta function equal to ∞ when $\mathbf{r} = \mathbf{r}'$ and 0 otherwise. Thus

$$\int_{\text{direct}} \delta(\mathbf{r} - \mathbf{r}') d\mathbf{r}'^3 = 1 \qquad (3.25)$$

In other words, $\delta(\mathbf{r} - \mathbf{r}')$ represents an infinitely sharp line of unit weight at $\mathbf{r} = \mathbf{r}'$. Hence, we have shown that

$$G(\mathbf{r}') = \int_{\text{direct}} \rho(\mathbf{r}')\delta(\mathbf{r} - \mathbf{r}') d\mathbf{r}'^3 = \rho(\mathbf{r}) \qquad (3.26)$$

We see therefore that

$$A(\mathbf{Q}) = \int_{\text{direct}} \rho(\mathbf{r}) e^{i\mathbf{r} \cdot \mathbf{Q}} d\mathbf{r}^3 \qquad (3.27)$$

and

$$\rho(\mathbf{r}) = \int_{\text{Fourier}} A(\mathbf{Q}) e^{-i\mathbf{r} \cdot \mathbf{Q}} d\mathbf{Q}^3 \qquad (3.28)$$

This means that we can convert forwards and backwards between direct and Fourier, i.e., reciprocal space, with these two equations.

Note: *In terms of lattice theory the reciprocal lattice is the Fourier transform of the real lattice and the real lattice is the Fourier transform of the reciprocal lattice.*

3.7 Convolution theorem

If we wish to gain some insight into diffraction calculations for a particular crystal structure it is quite instructive to go back to the way we earlier defined the crystal structure $C(\mathbf{r})$ in terms of a convolution of a lattice function $L(\mathbf{r})$

$$L(\mathbf{r}) = \sum_{uvw} \delta(\mathbf{r} - \mathbf{r}_{uvw}) \qquad (3.29)$$

and a basis $B(\mathbf{r})$:

$$C(\mathbf{r}) = L(\mathbf{r}) * B(\mathbf{r}) \qquad (3.30)$$

Now to calculate the diffraction information from something that is described by the convolution of two functions we can make good use of the so-called convolution theorem. Consider the convolution of two functions $f(\mathbf{r})$ and $g(\mathbf{r})$:

$$h(\mathbf{r}) = f(\mathbf{r}) * g(\mathbf{r}) \qquad (3.31)$$

The Fourier transform $H(\mathbf{Q})$ of $h(\mathbf{r})$ is given by

$$H(\mathbf{Q}) = \int h(\mathbf{r}) e^{i\mathbf{r} \cdot \mathbf{Q}} d\mathbf{r}^3 \qquad (3.32)$$

and then

$$H(\mathbf{Q}) = \int e^{i\mathbf{r}\cdot\mathbf{Q}}[\int f(\mathbf{r}')g(\mathbf{r}-\mathbf{r}')d\mathbf{r}'^3]d\mathbf{r}^3 \quad (3.33)$$

Changing the order of the integration

$$H(\mathbf{Q}) = \int f(\mathbf{r}')[\int e^{i\mathbf{r}\cdot\mathbf{Q}}g(\mathbf{r}-\mathbf{r}')d\mathbf{r}^3]d\mathbf{r}'^3 \quad (3.34)$$

We now make the substitution $\mathbf{r} - \mathbf{r}' = \mathbf{R}$, and then $d\mathbf{r} = d\mathbf{R}$

$$H(\mathbf{Q}) = \int f(\mathbf{r}')[\int e^{i(\mathbf{R}+\mathbf{r}')\cdot\mathbf{Q}}g(\mathbf{R})d\mathbf{R}^3]d\mathbf{r}'^3 \quad (3.35)$$

and this becomes

$$H(\mathbf{Q}) = \int f(\mathbf{r}')e^{i\mathbf{r}'\cdot\mathbf{Q}}d\mathbf{r}'^3 \int g(\mathbf{R})e^{i\mathbf{R}\cdot\mathbf{Q}}d\mathbf{R}^3 \quad (3.36)$$

In other words,

$$f(\mathbf{r}) * g(\mathbf{r}) \stackrel{FT}{\longleftrightarrow} F(\mathbf{Q}) \times G(\mathbf{Q}) \quad (3.37)$$

where $F(\mathbf{Q})$ and $G(\mathbf{Q})$ are the Fourier transforms of the functions $f(\mathbf{r})$ and $g(\mathbf{r})$, respectively. Conversely

$$f(\mathbf{r}) \times g(\mathbf{r}) \stackrel{FT}{\longleftrightarrow} F(\mathbf{Q}) * G(\mathbf{Q}) \quad (3.38)$$

Theorem: *The convolution theorem states that the Fourier transform of the convolution of two functions is equal to the product of the individual Fourier transforms. Conversely, the Fourier transform of the product of two functions is equal to the convolution of the individual Fourier transforms.*

Applying this to a crystal structure, the Fourier transform relationship is given symbolically by

$$\tilde{C}(\mathbf{r}) = \tilde{L}(\mathbf{r}) \times \tilde{B}(\mathbf{r}) \quad (3.39)$$

where the Fourier transforms are denoted by the 'squiggles'.

We can go even farther with this idea. Suppose that the crystal is actually finite in size. We then define a shape function $S(\mathbf{r})$ that can be stated as follows

$$\begin{aligned} S(\mathbf{r}) &= 1 \quad \text{for} \quad r \leqslant r_0 \\ &= 0 \quad \text{for} \quad r > r_0 \end{aligned} \quad (3.40)$$

r_0 represents the boundary distance of the crystal. So, a finite crystal can be defined by

$$C(\mathbf{r}) = [S(\mathbf{r}) \times L(\mathbf{r})] * B(\mathbf{r}) \quad (3.41)$$

Note that the shape function multiplies the lattice function $L(\mathbf{r})$ thus ensuring that there are no lattice points outside the region defined by the shape function.

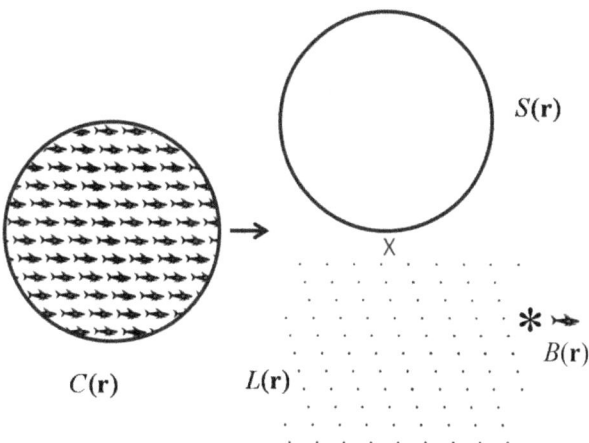

Figure 3.23. A finite crystal in terms of a shape function, a molecule (in the shape of a fish here!) and a lattice.

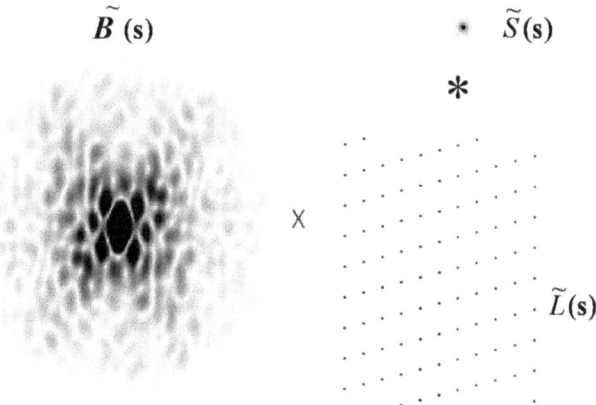

Figure 3.24. Fourier transformation of a finite crystal.

Figure 3.23 illustrates the way in which a finite crystal is built up of these three elements. The Fourier transform then becomes (figure 3.24).

$$\tilde{C}(\mathbf{Q}) = [\tilde{S}(\mathbf{Q}) * \tilde{L}(\mathbf{Q})] \times \tilde{B}(\mathbf{Q}) \tag{3.42}$$

Note the way in which the large shape function becomes much smaller in reciprocal space.

Finally, in figure 3.25 the diffraction pattern is shown. Notice that all the reciprocal lattice points are no longer described by delta functions but are extended because of the convolution of the shape function with the reciprocal lattice.

Furthermore, multiplication by the molecular transform results in different peaks at the reciprocal lattice nodes: the underlying molecular transform is sampled at the reciprocal lattice points to give amplitudes (and hence intensities) of differing height. Clearly for a large crystal, the shape function will become much smaller and so the reflections will become much narrower in extent.

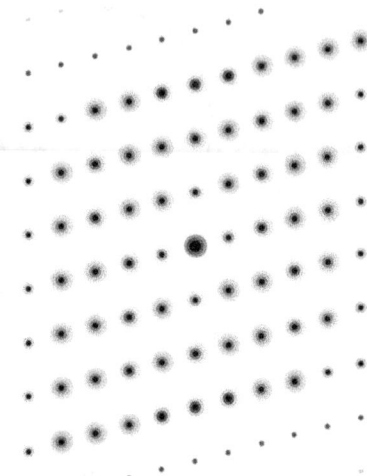

Figure 3.25. Intensity transform of finite crystal.

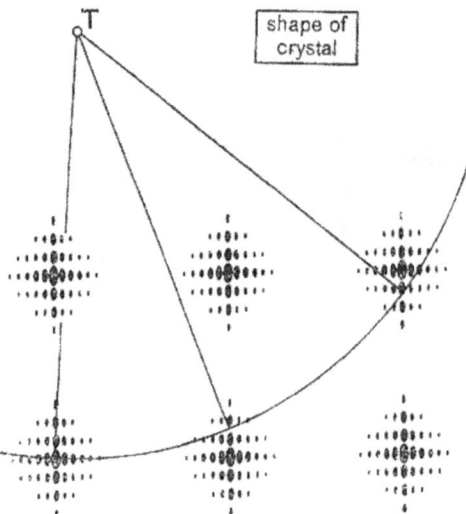

Figure 3.26. Intensity distribution surrounding reciprocal lattice points for a finite rectangular crystal. Reproduced from [5] with permission of Springer.

For very small, perfect crystals the Fourier transform of the shape function will have a central maximum surrounded by weaker intensity subsidiary maxima. Intersection with the Ewald sphere (figure 3.26) will therefore lead to more complex intensity distributions around the reciprocal lattice nodes. As the crystal is made larger these maxima will move into the central maxim which will shrink in width. However, at the same time the effect of the increased crystal volume will lead to enlargement of the peak widths, thus counteracting the diffraction effect.

Other examples are shown in figure 3.27. On the top are shown three masks in which holes have been made to represent atoms. Below are the diffraction patterns

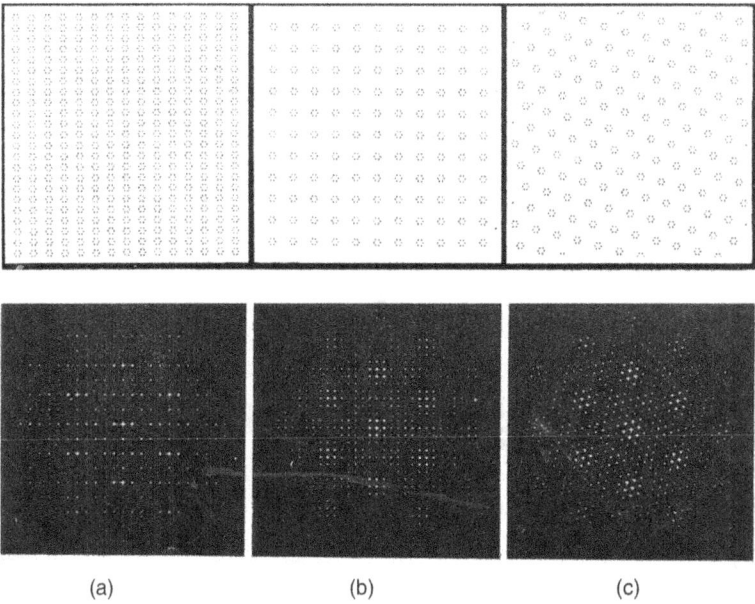

Figure 3.27. Optical transforms of a crystal structure with a molecule consisting of a hexagonal ring of atoms [6].

produced by diffraction of light through the masks. It can be seen in (a) that the structure consists of hexagonal molecules stacked together in a repeating pattern according to a real orthogonal lattice. The diffraction pattern below shows sharp spots of varying intensities, just as one would expect from diffraction of x-rays by a real crystal. In (b) the intermolecular distances have been increased. The effect of this is to create a reciprocal lattice with points closer together. It can now be seen in the diffraction pattern below how the reciprocal lattice samples the underlying transform of the molecule itself (notice the 6-fold symmetry). In (c) the molecules have been repeated on a hexagonal net, and its diffraction pattern again shows the underlying molecular transform but the pattern of reciprocal lattice points now has hexagonal symmetry.

In figure 3.28 the effect of crystal shape is again demonstrated. (a) shows the diffraction pattern from a simple regular array of atoms. Note that the reciprocal lattice points have become spread vertically and horizontally: this is because the structure shown above is not infinite in extent but is bounded by the square outline of the figure. The Fourier transform of a square will be elongated in directions perpendicular to the edges of the square and contracted in the directions along the diagonals (dimensions in reciprocal space are inverted with respect to real space). In (b) the horizontal dimension of the 'crystal' has been reduced and the transform shows corresponding horizontal spreading of the intensity at each reciprocal lattice node. In (c) the crystal is rotated through an angle and we see that the spread around each node has also rotated.

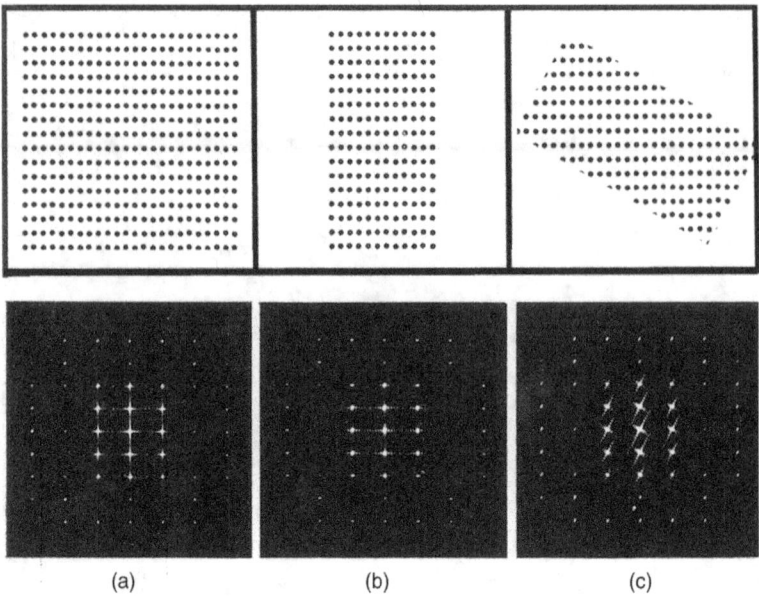

Figure 3.28. Optical transforms of a crystal structure with an array of atoms [6].

3.8 Two simple 'Rules'

I find it useful when teaching students about diffraction to visualize Fourier transformation according to two simple rules:
1. Fundamental symmetry is preserved.
2. Large distances become small and small distances large.

Although this is a gross simplification, nonetheless by remembering this we can check that a calculated Fourier transformation makes sense. Let us look at a couple of examples.

In figure 3.29 a one-dimensional vertical lattice L_1 is shown and on the right a schematic view of its Fourier transform. First of all, the lattice has translational symmetry along the c-axis and so this symmetry will be preserved in the Fourier transform (Rule 1). However, the spacing of the lattice points will become inverted (Rule 2). Furthermore, the dimension of the one-dimensional lattice in the horizontal plane is zero and so in Fourier space (Rule 2) it will be extended infinitely. The result is a series of horizontal infinite planes separated along c with translational symmetry (Rule 1).

Now consider a two-dimensional lattice (figure 3.30). This can be built from a convolution of two one-dimensional lattices L_1 and L_2. The Fourier transform therefore, according to the convolution theorem, will be given by $\tilde{L}_1 \times \tilde{L}_2$. This means that in Fourier space there will be one set of parallel planes perpendicular to L_1 and one set perpendicular to L_2 whose spacings are proportional to the reciprocal of the real spacings of the lattice points. Because of the multiplication of the two

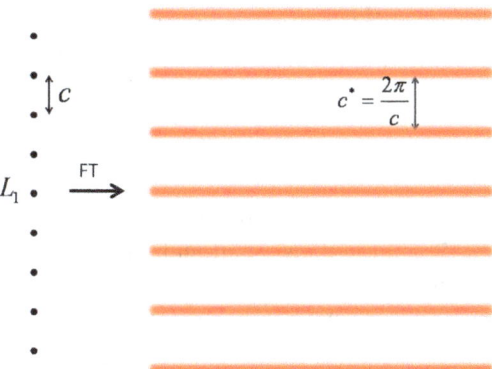

Figure 3.29. Fourier transformation of a one-dimensional lattice. The horizontal lines are in fact planes perpendicular to c.

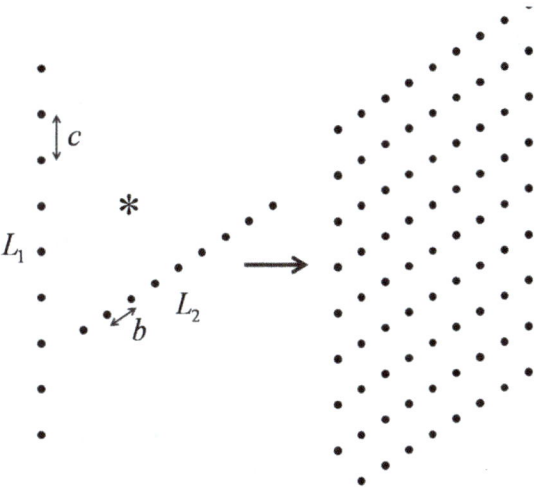

Figure 3.30. Convolution of two one-dimensional lattices.

Fourier transforms everything will cancel except where the planes cross to give infinite lines perpendicular to $b*c*$ (Rule 2).

It should be obvious that if we introduce a third one-dimensional lattice L_3 the Fourier transform will then be $\tilde{L}_1 \times \tilde{L}_2 \times \tilde{L}_3$. The three sets of infinite lines produced from the Fourier transforms of the lattice functions will cross, giving zero everywhere, except where they intersect, to create a three-dimensional lattice of points—the reciprocal lattice again!

3.9 Lattice diffraction

Let us return now to the phase difference produced when waves are scattered by two different points a distance r apart. We shall assume here that the so-called first Born approximation applies, namely that only a small fraction of the incident radiation is actually scattered. This is also known as the kinematic approximation in diffraction theory. We also assume that the scattering is elastic, i.e., there is no difference in energy between incident and diffracted waves. To understand the intensity observed in a normal diffraction pattern, we need also to realize that the observed diffraction pattern can be considered to be effectively at a very large distance from the scattering object, and so we can treat this as a case of Fraunhofer, rather than of Fresnel, diffraction.

We start by imagining that the lattice alone is involved in the diffraction process. Note that, though the lattice is actually a fictional quantity and therefore has no real existence, we can nonetheless from a mathematical point of view imagine that it consists of an infinite regular array of points capable of diffracting incident waves. As in equation (3.1), the path difference will then be given by

$$\mathbf{r} \cdot (\mathbf{s} - \mathbf{s}_0) \quad (3.43)$$

and so the phase difference ϕ between the waves will be

$$\phi = \frac{2\pi}{\lambda} \mathbf{r} \cdot (\mathbf{s} - \mathbf{s}_0) = \mathbf{r} \cdot (\mathbf{k} - \mathbf{k}_0) = \mathbf{r} \cdot \mathbf{Q} \quad (3.44)$$

The vector $\mathbf{Q}$ is the scattering vector measured from the origin of reciprocal space to anywhere else in reciprocal space, i.e., in general it is not necessarily confined to ending at reciprocal lattice points but can terminate between reciprocal lattice points to sample any background diffraction intensity between the reflections. The amplitude scattered by the array of points will then be given by an integral over the volume space $\mathbf{r}^3$ spanned by the lattice:

$$A(\mathbf{Q}) = \int_{\text{lattice}} e^{i\mathbf{r} \cdot \mathbf{Q}} \, d\mathbf{r}^3 \quad (3.45)$$

Now, let us assume an infinite lattice (a pretty good approximation in general) and consider the observed intensity located only at the nodes of the reciprocal lattice. From equations (3.4) and (3.11) at the reciprocal lattice nodes the scattering vector $\mathbf{Q}$ will be equivalent to the reciprocal lattice vector $\mathbf{g}_{hkl}$. Also

$$\mathbf{r} = n_1 \mathbf{a} + n_1 \mathbf{b} + n_1 \mathbf{c} \quad (3.46)$$

where n_1, n_2 and n_3 are integers specifying all the real primitive lattice nodes. Therefore

$$\begin{aligned} \mathbf{r} \cdot \mathbf{Q} &= \mathbf{r} \cdot \mathbf{g}_{hkl} \\ &= (n_1 \mathbf{a} + n_2 \mathbf{b} + n_3 \mathbf{c}) \cdot (h\mathbf{a}^* + k\mathbf{b}^* + l\mathbf{c}^*) \\ &= 2\pi(n_1 h + n_2 k + n_3 l) \end{aligned} \quad (3.47)$$

and then

$$A(\mathbf{Q}) = \int_{\text{lattice}} e^{i 2\pi (n_1 h + n_2 k + n_3 l)} d\mathbf{r}^3 \quad (3.48)$$

and this only has values at the discrete points hkl of the reciprocal lattice described by the reciprocal lattice vectors. So, we can replace the integral by a summation over all lattice points:

$$F(hkl) = \sum_{n_1 n_2 n_3} e^{2\pi i(n_1 h + n_2 k + n_3 l)} \quad (3.49)$$

where we write $F(hkl) = A(\mathbf{Q})$ for amplitudes observed only at reciprocal lattice points.

Note: *We again see from this that in fact the reciprocal lattice can be thought of as the Fourier transform of the real lattice and vice versa.*

The intensities at the points *hkl* are given by

$$I(hkl) \propto F(hkl)F^*(hkl) = |F(hkl)|^2 \quad (3.50)$$

and this shows that the observed array of diffraction spots can be thought of as a kind of intensity 'imprint' of the reciprocal lattice.

3.10 Structure factors

We now turn to calculation of diffraction amplitudes and intensities in terms of the crystal structure. As usual, the amplitude $A(\mathbf{Q})$ anywhere in reciprocal space is given by equation (3.20) where $\rho(\mathbf{r})$ is the volume density of the scattering material. In x-ray diffraction, the scattering is due to the electron distribution in the crystal and hence $\rho(\mathbf{r})$ is the electron density (number of electrons per unit volume) at any position $\mathbf{r}$ in the crystal.

Now let us introduce the fact that the crystal structure is periodic, and so we can sum over *j* atoms at distances $\mathbf{r}_j$ from a chosen origin of a unit cell. Referring to figure 3.31 the integral then becomes

$$\begin{aligned}
A(\mathbf{Q}) &= \int \sum_j \rho_j(\mathbf{r} - \mathbf{r}_j) e^{i\mathbf{r} \cdot \mathbf{Q}} \, d\mathbf{r}^3 \\
&= \sum_j e^{i\mathbf{r} \cdot \mathbf{Q}} \int \rho_j(\mathbf{r} - \mathbf{r}_j) e^{i(\mathbf{r} - \mathbf{r}_j) \cdot \mathbf{Q}} \, d(\mathbf{r} - \mathbf{r}_j)^3 \quad (3.51) \\
&= \sum_j e^{i\mathbf{r} \cdot \mathbf{Q}} \int \rho_j(\mathbf{r}') e^{i\mathbf{r}' \cdot \mathbf{Q}} \, d\mathbf{r}'^3
\end{aligned}$$

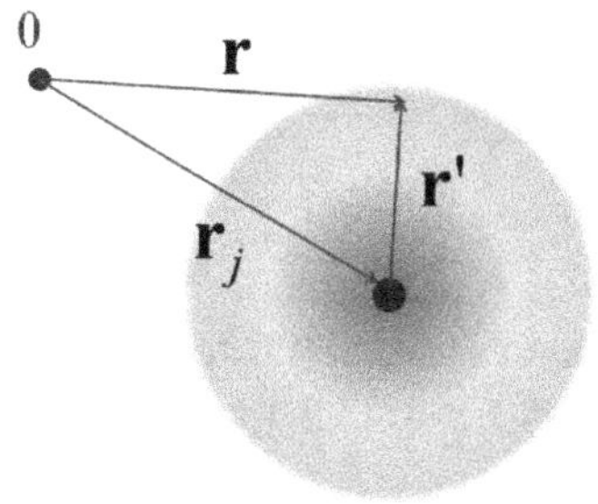

Figure 3.31. Scattering of x-rays by electron density around an atomic nucleus.

Now, for an infinite crystal, the observed intensity will be located at the nodes of the reciprocal lattice. Therefore once again from equations (3.4) and (3.11)

$$\mathbf{Q} = \mathbf{g}_{hkl} \tag{3.52}$$

and

$$\mathbf{r}_j = x_j\mathbf{a} + y_j\mathbf{b} + z_j\mathbf{c} \tag{3.53}$$

where x_j, y_j, z_j, are the fractional coordinates of the jth atom expressed as a fraction of the unit cell axial lengths. Therefore

$$\begin{aligned}\mathbf{r}_j \cdot \mathbf{Q} = \mathbf{r}_j \cdot \mathbf{g}_{hkl} &= (x_j\mathbf{a} + y_j\mathbf{b} + z_j\mathbf{c}) \cdot (h\mathbf{a}^* + k\mathbf{b}^* + l\mathbf{c}^*) \\ &= 2\pi(hx_j + ky_j + lz_j) \end{aligned} \tag{3.54}$$

Writing

$$f_j = \int \rho_j(\mathbf{r}') e^{i\mathbf{r}' \cdot \mathbf{Q}} \, d\mathbf{r}'^3 \tag{3.55}$$

the resulting scattered amplitude **F(hkl)** is then given by

$$\mathbf{F}(hkl) = \sum_j f_j e^{2\pi i(hx_j + ky_j + lz_j)} \tag{3.56}$$

Definition *The quantity* **F(h k l)** *is known as the* **structure factor** *and its modulus* |**F(h k l**)| *is the* **structure amplitude**. f_j *is called the* **atomic scattering factor** *or* **form factor**.

3.11 Form factors

Now I know what the atom looks like.

Ernest Rutherford

X-rays

Let the vector $\mathbf{r}'$ make an angle Ω with the vector $\mathbf{Q}$. Then

$$\mathbf{r}' \cdot \mathbf{Q} = r'Q \cos \Omega \tag{3.57}$$

The quantity ρ_j is usually taken to be the electron density of atom j and is assumed to be spherically distributed about the atomic nucleus. This is generally a good assumption for most purposes but does not take into account the much weaker effect of the electron clouds being distorted through bonding to neighbouring atoms. Therefore from equation (3.55) with a spherical density distribution about the nucleus:

$$f_j = 2\pi \int \int \rho_j(\mathbf{r}')r'^2 e^{ir'Q\cos\Omega}\,dr'\,d(\cos\Omega)$$
$$= 2\pi \int \rho_j(\mathbf{r}')r'^2 \frac{e^{ir'Q} - e^{-ir'Q}}{ir'Q}\,dr' \qquad (3.58)$$

and then

$$f_j = 4\pi \int \rho_j(\mathbf{r}')r'^2 \frac{\sin r'Q}{r'Q}\,dr' \qquad (3.59)$$

For free atoms, the electronic wave-functions ψ_j can be calculated to good accuracy, and therefore the electron densities $|\psi_j\psi_j^*|$ can also be evaluated and the atomic scattering factors computed and tabulated. Thus, for x-rays scattered in the forward direction ($Q = 0$)

$$f_j = 4\pi \int \rho_j(r')r'^2\,dr' = Z_j \qquad (3.60)$$

with Z_j the atomic number. As the angle of diffraction 2θ increases, the diffracted waves from the electrons will destructively interfere, thus reducing the scattered amplitude. The atomic scattering factors can be found in standard tables, where they are usually given as a function of $\sin\theta/\lambda$.

Note how in figure 3.32 the atomic scattering factors drop smoothly from their Z values as the angle of scattering increases. You can also see in this figure that the ions K^+ and Cl^-, which are isoelectronic, have the same values at $\theta = 0$ but differ slightly as the angle of scattering increases.

Using the two Fourier transform rules we can understand why the atomic scattering factor looks like this as a function of $\sin\theta/\lambda$. The electron density of the

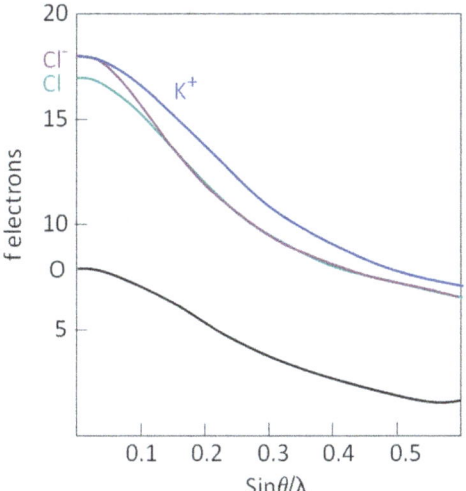

Figure 3.32. Examples of x-ray atomic scattering factors plotted as a function of $\sin\theta/\lambda$.

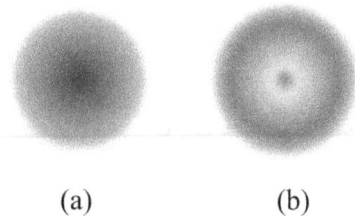

(a) (b)

Figure 3.33. Simulation of diffraction from (a) a gas and (b) a liquid or amorphous material.

atom j is taken to be spherically distributed, and so Rule 1 means that f_j will be isotropic in space. Rule 2 means that f_j will drop off with angle faster for heavier, and thus larger, atoms (more electrons) than for lighter atoms, just as is seen in figure 3.32.

The atomic scattering factor can be used on its own to calculate the intensity of diffraction of x-rays by a gas of atoms. In this case the intensity is simply given by

$$I(\mathbf{Q}) \propto [\sum f_j]^2 \tag{3.61}$$

Note that in this case the exponential term in the structure factor formula is removed, since in a gas the atoms are randomly distributed throughout space and no lattice function applies. The scattering vector $\mathbf{Q}$ now spans all reciprocal space. Because a gas is isotropic, the resulting diffraction pattern consists of a spherical distribution of scattered rays whose intensity drops off with angle.

Figure 3.33(a) shows a simulation of approximately what the x-ray diffraction pattern of a gas (no long- or short-range order present) would look like (not very interesting). In figure 3.33(b) the effect of diffracting x-rays from a liquid or an amorphous material (i.e. a substance with no long-range order of its atomic structure) is shown. Here we see some diffuse rings resulting from the fact that in a condensed state, such as in a liquid or amorphous substance, atoms will come into near contact and therefore there will always be some short-range order present.

3.12 Anomalous dispersion

In discussing the atomic scattering factor for x-rays, we need to consider what happens when x-rays are strongly absorbed by the electrons in an atom. The effect is to cause the electrons to undergo inter-shell transitions at certain energies.

Consider the formula for the standard damped harmonic motion of a classical electron in an oscillating electric field $E_0 e^{i\omega t}$:

$$\ddot{x} + \gamma \dot{x} + \omega_0^2 = \frac{e}{m_e} E_0 e^{i\omega t} \tag{3.62}$$

where γ is a damping constant and ω_0 is the resonant frequency of the bound electron. If we take the solution to be of the form

$$x = x_0 e^{i\omega t} \tag{3.63}$$

we then get

$$x = \frac{e}{m_e}\left(\frac{E_0 e^{i\omega t}}{\omega_0^2 - \omega^2 + i\gamma\omega}\right) \quad (3.64)$$

x then characterizes an oscillating electron dipole which acts as a source of scattered electromagnetic radiation. At a large distance R, the wave can be thought of as spherical and its electric vector in the plane perpendicular to the dipole is $\omega^2/c^2|\mathbf{R}|$ times the dipole moment ex at the time $t - |\mathbf{R}|/c$. Then the amplitude of the scattered wave is

$$\frac{e^2}{m_e c^2}\left(\frac{\omega^2 E_0}{\omega_0^2 - \omega^2 + i\gamma\omega}\right) \quad (3.65)$$

The atomic scattering factor f is given by the ratio of the amplitude scattered by the oscillator to that scattered by a free classical electron, where $\omega_0 = 0$ and $\gamma = 0$. The amplitude for the free electron therefore is

$$-\frac{e^2}{m_e c^2} E_0 \quad (3.66)$$

which means that the wave scattered by the free electron in the forward direction has opposite phase to that of the incident wave. Therefore, we find that for a positive value of f

$$f = \frac{\omega^2}{\omega_0^2 - \omega^2 - i\gamma\omega} \quad (3.67)$$

We can the generalize the formula for the atomic scattering factor as

$$f = f_0 + \Delta f' + \Delta f'' \quad (3.68)$$

where f_0 is the Fourier transform of the electron distribution in the atom. Figure 3.34 shows a theoretical plot of the real and imaginary anomalous dispersion terms as a function of wavelength for the metal platinum.

Provided that one is not close to one of the dipole resonances, the second and third terms are relatively small and can be ignored. In this case, consider the amplitudes for a reflection $h\ k\ l$ and for a reflection $\bar{h}\bar{k}\bar{l}$:

$$\mathbf{F}(hkl) = \sum f_{j0} e^{2\pi i(hx_j + ky_j + lz_j)} \quad (3.69)$$

and

$$\mathbf{F}(\bar{h}\bar{k}\bar{l}) = \sum f_{j0} e^{-2\pi i(hx_j + ky_j + lz_j)} \quad (3.70)$$

But the intensity is given by the square of the modulus of the structure factor, whence

$$I(hkl) \propto \mathbf{F}(hkl)\mathbf{F}^*(hkl) = \mathbf{F}(\bar{h}\bar{k}\bar{l})\mathbf{F}^*(\bar{h}\bar{k}\bar{l}) \quad (3.71)$$

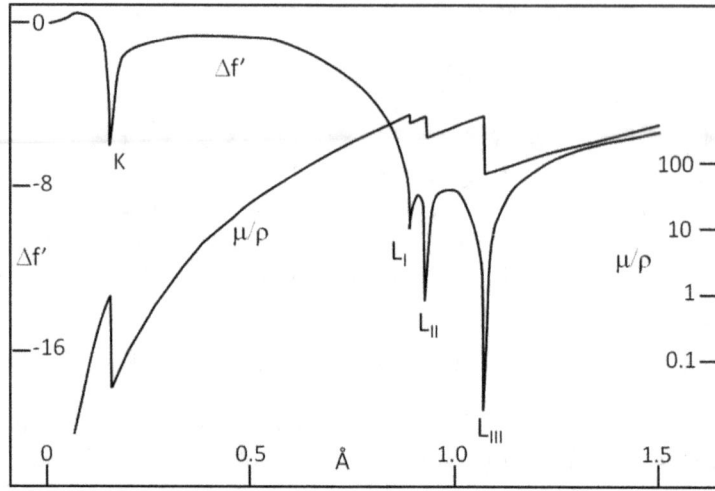

Figure 3.34. Theoretical x-ray dispersion corrections for platinum. $\Delta f'$ is effectively measured in numbers of electrons. μ/ρ is the mass absorption coefficient in $cm^2\ g^{-1}$ and is proportional to $\Delta f''$.

Definition: In *the absence of x-ray absorption $I(hkl) = I(\bar{h}\bar{k}\bar{l})$ and the diffraction pattern appears to be centrosymmetric. This is known as **Friedel's Law.***

However, when working with x-rays close to the absorption edge for a particular atom, the extra terms do become important and because of the imaginary term $\Delta f''$ Friedel's Law no longer holds true. This departure from centrosymmetry of the diffraction pattern can be used for instance to determine in a structurally polar crystal, such as a ferroelectric, which way certain atoms are displaced in order to link this with the sign of the macroscopic polarization. Similarly, it can be used to differentiate between two isostructural optically active chiral materials (i.e. those that rotate the plane of polarized light either to the right or to the left). In 1951, Johannes Martin Bijvoet and his research group used this anomalous dispersion effect to determine successfully for the first time the absolute configuration of an optically active crystal, in this case sodium rubidium tartrate. Anomalous dispersion with multiple wavelengths (multi-wavelength anomalous dispersion, MAD) in more recent times has been used to solve the phase problem for complex protein structures that have thousands of atoms in a unit cell.

Neutrons

The first evidence for the diffraction of neutrons was made in 1936 by Mitchell and Powers [7] and by von Halban and Preiswerk [8]. But it was Bertram Brockhouse and Clifford Shull who developed neutron spectroscopy and diffraction in the 1940s, and who shared the Nobel Prize for this work as late as 1994.

Table 3.1. Neutron scattering lengths and cross sections.

Isotope	$b_{coh}(fm)$	$b_{inc}(fm)$	$\sigma_{coh}(barns)$	$\sigma_{Inc.}(barns)$
1_H	−3.74	25.27	1.76	80.27
2_H	6.67	4.04	5.59	2.05
16_O	5.80	0	4.23	0
39_K	3.74	1.4	1.76	0.25
35_{Cl}	11.65	6.1	17.06	4.7

In neutron diffraction, the scattering objects are the atomic nuclei. The neutron of mass m_N travels with a velocity v_N and then according to the de Broglie relationship can be described by a wave of wavelength given by

$$\lambda = \frac{h}{m_N v_N} \qquad (3.72)$$

which for thermal neutrons is of the same order as the x-ray wavelengths used in diffraction, despite the fact that their energies are very different. For example, a wavelength of approximately 1 Å corresponds to a neutron energy of 0.1 eV, to be compared with x-rays with energy 12.5 keV. The formalism is essentially the same, except that in the structure factor formula the form factor is conventionally given by b, the so-called *scattering length* or σ, the scattering *cross-section*. The scattering length is measured in units of femtometres (1×10^{-15} m) and the cross-section in barns (1×10^{-24} cm^2).

Now if we follow the Fourier transform rules, the scattering will again be isotropic, as we can assume that the scattering object, the nucleus, is spherically symmetric. The nucleus is extremely small compared with the electron cloud around the nucleus, so that in Fourier space the observed amplitude of nuclear scattering will persist to very large angles. In fact, so much so that it can be taken to be independent of angle. Table 3.1 gives a few example values.

In table 3.1, both coherent and incoherent scattering components are given: the coherent terms are those that give rise to sharp Bragg reflections, whilst the incoherent terms contribute to background scattering. Because the mass of a ^{1}H nucleus is almost the same as that of a neutron it can be displaced by collision with the incoming neutron and this results in a large incoherent cross-section. Therefore, neutron diffraction from a crystal containing substantial amounts of hydrogen can be compromised by creating too much background scattering. It is for this reason that neutron diffraction studies of hydrogen-containing crystals are often done with the hydrogen nuclei substituted by deuterium ^{2}H for which the incoherent scattering length is much smaller. Note also that the nuclear scattering length can be negative, unlike x-ray form factors, signifying that the outgoing wave has the sign of its phase reversed compared with the equivalent x-ray wave. It can also be seen that ^{39}K and ^{35}Cl nuclei have quite different scattering lengths, so that with neutron diffraction it is easy to distinguish between the effects of these nuclei. Contrast this with the case of x-ray scattering where the atomic scattering factors of K$^+$ and Cl$^-$ ions are almost identical. So, for instance an x-ray powder pattern of KCl will look like it is primitive cubic, since the K$^+$ and Cl$^-$ ions will be almost indistinguishable from one another.

There is also another type of neutron scattering that needs to be considered. The neutron possesses spin and this means it can also interact with magnetic moments, such as those due to unpaired electrons in an atom. This gives rise to an additional diffraction pattern superimposed upon the nuclear diffraction pattern. Note that in magnetic scattering the scattering object is the cloud of electrons around a nucleus, and so, as in x-ray diffraction, its form factor falls off rapidly with angle. The effect of magnetic neutron scattering can be particularly easily seen for antiferromagnetic crystals where the magnetic moment may alternate in direction, thus cancelling out the net magnetization. In such cases the magnetic moments repeat at distances as a multiple, say double, of the repeat distance of the nuclei, thus doubling the unit nuclear unit cell. This leads to reciprocal lattice points appearing half-way between those due to the nuclear scattering.

Electrons

The diffraction of electrons by atoms in crystals was discovered by Clinton Davisson in the USA and by George Paget Thomson (son of J J Thomson the discoverer of the electron)[2] who shared the Nobel Prize in 1937. A good book on electron diffraction is by Grundy and Jones [9].

In electron diffraction, the scattering is via the electrons in the solid distributed according to a periodic potential energy $V(\mathbf{r})$. Assuming spherically symmetric atoms, elastic scattering and the first Born approximation with incident electron energy $E \gg V(\mathbf{r})$

$$f^{el} = \frac{m_e}{2\pi\hbar^2} \int V(\mathbf{r}) e^{i\mathbf{r}\cdot\mathbf{Q}} \, dV(\mathbf{r})$$
$$= \frac{m_e}{2\pi\hbar^2 Q^2} \int V(\mathbf{r}) \nabla^2 (e^{i\mathbf{r}\cdot\mathbf{Q}}) \, dV(\mathbf{r}) \quad (3.73)$$

Integrating by parts and using $V(\infty) = V(-\infty) = 0$

$$f^{el} = \frac{m_e}{2\pi\hbar^2 Q^2} \int e^{i\mathbf{r}\cdot\mathbf{Q}} \nabla^2 V \, dV(\mathbf{r}) \quad (3.74)$$

Using Poisson's equation relating the potential and charge distributions

$$\nabla^2 V(\mathbf{r}) = \frac{e[Z\delta(\mathbf{r}) - \rho(\mathbf{r})]}{\varepsilon_0} \quad (3.75)$$

where $Z\delta(\mathbf{r})$ is the nuclear point charge and ε_0 is the permittivity of free space. Therefore

---

[2] Thus, the father identified electrons as particles while his son showed they were waves!

$$f^{el} = \frac{m_e e}{2\pi\hbar^2\varepsilon_0 Q^2}\int [Z\delta(\mathbf{r}) - \rho(\mathbf{r})]e^{i\mathbf{r}\cdot\mathbf{Q}}\,dV(\mathbf{r}) \tag{3.76}$$

We then obtain the Mott–Bethe formula:

$$f^{el} = \frac{m_e e}{2\pi\hbar^2\varepsilon_0 Q^2}[Z - f^X] \tag{3.77}$$

where f^X is the x-ray form factor. Typically, f^{el} is in the range 1–10 Å, whereas f^X is of the order of 10^{-4} Å, and so electrons are scattered more strongly than x-rays (or neutrons). The angular range for electron scattering is therefore very small, and hence the scattering is strongly peaked in the forward direction. For 100 keV electrons

$$\lambda = \frac{h}{\sqrt{2m_e V}} = \frac{12.6}{\sqrt{V}} \simeq 0.04 \text{ Å} \tag{3.78}$$

and therefore,

$$\Delta(\sin\theta) \simeq 0.02 \tag{3.79}$$

and

$$\Delta\theta \simeq 2\times 10^{-2}\,\text{rad} \simeq 1° \tag{3.80}$$

We can also see this in terms of Fourier transforms. The periodic potential extends over a large range so that using the rules, the scattered electrons will be confined to a very narrow angular range about the origin. The typical voltages (wavelengths) used in electron diffraction are shown in table 3.2 where LEED means low-energy electron diffraction, MEED medium-energy electron diffraction and HEED high-energy electron diffraction. v/c is the fraction of the electron velocity compared with the velocity of light. For excitation voltages less than about 100 V the electrons are

Table 3.2. Typical voltages used in electron diffraction

V (volts)	λ(Å)	v/c	Name
10	3.9	0.006	LEED
10^2	1.2	0.02	LEED
10^3	0.39	0.06	MEED
10^4	0.12	0.19	MEED
10^5	0.037*	0.54	HEED
10^6	0.009*	0.94	HEED
	*relativistic correction of ~5% at 10^5 V		

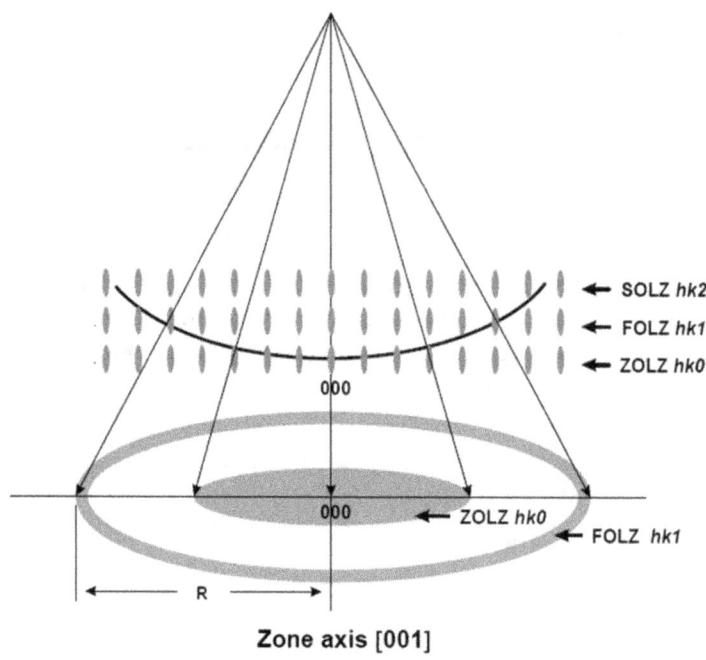

Figure 3.35. Ewald sphere construction in transmission electron microscopy (TEM). Here the c-axis of the crystal is perpendicular to the crystal plate. ZOLZ = zero-order Laue zone; FOLZ = first-order Laue zone; SOLZ = second-order Laue zone (from [10]).

very strongly absorbed, and so they only penetrate a few atomic layers of a crystal. Thus, LEED tends to be used for surface studies, whereas standard transmission electron diffraction and microscopy uses voltages in excess of 10^5 V. In transmission electron diffraction, because of the high absorption, very thin crystals are normally used.

It is instructive to consider the reciprocal lattice and the Ewald sphere construction in this case (figure 3.35). In the electron microscope, a thin sample is placed at the sample-stage position with a perpendicular incident beam of electrons. Now because the wavelength of the electron beam is much smaller than that normally used with x-rays, the radius of the corresponding Ewald sphere is very large, making its surface almost flat. Because the crystal is thin, then according to the Fourier transform rules, the reciprocal lattice points will be extended to form truncated rods in a direction perpendicular to the crystal plane. The surface of the Ewald sphere then cuts through many reciprocal lattice rods. The result is that even though the crystal is stationary, the diffraction image shows many reflections simultaneously. In figure 3.36, the spots observed near the centre are those for which the Ewald sphere intersects the rods in the $hk0$ layer while the outer ring of spots comes from the $hk1$ layer. These circular areas of diffraction are commonly known by electron diffractionists as *Laue zones*.

A Journey into Reciprocal Space

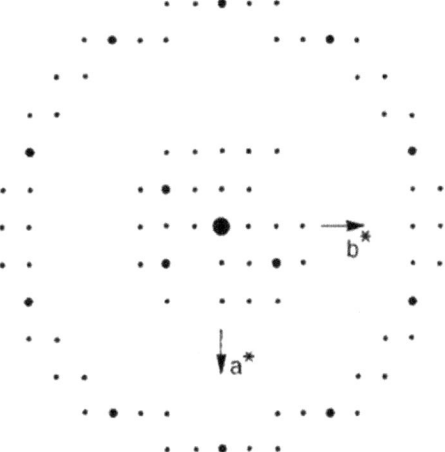

Figure 3.36. Electron diffraction image from a thin crystal.

3.13 Intensity calculations

For the purpose of explaining the principles behind the calculation of diffracted intensities, we shall adopt here a simplified approach, as is often used in undergraduate physics courses. We shall use the structure factors as a means of calculating the intensities of reflections.

As an example, consider the diamond crystal structure (figure 3.37) described in chapter 1. It would be straightforward to use the structure factor formula (3.56) and sum over all eight atoms. However, by making use of the convolution theorem we can simplify the process greatly using equation (3.39):

$$\mathbf{F}(hkl) = \left[f_C + f_C e^{i\pi\left(\frac{h+k+l}{2}\right)}\right][1 + e^{i\pi(h+k)} + e^{i\pi(h+l)} + e^{i\pi(k+l)}] \tag{3.81}$$

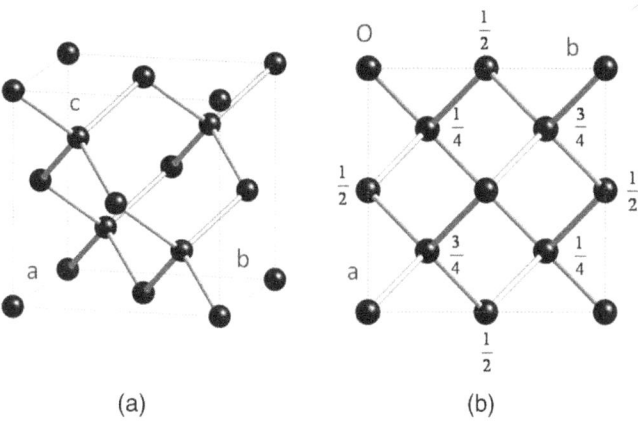

(a) (b)

Figure 3.37. The diamond crystal structure (a) normal view (b) projection onto the (001) plane.

3-38

Table 3.3. Lattice reflection conditions.

Lattice type	Reflection conditions
P	All hkl allowed
F	hkl all even or all odd
A	$k + l = 2n$
B	$h + l = 2n$
C	$h + k = 2n$
I	$h + k + l = 2n$
R-centred hexagonal	$-h + k + l = 3n$

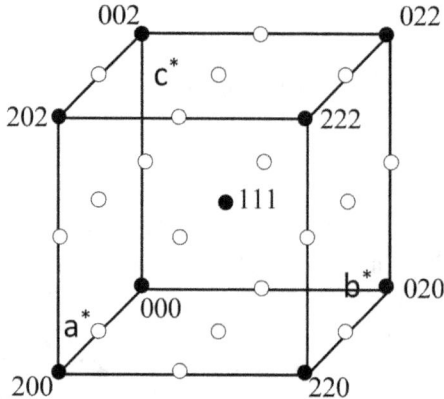

Figure 3.38. A reciprocal unit cell for an all-face-centred real lattice. Open circles denote points that are missing in the reciprocal lattice, where $h\ k\ l$ are not all even or all odd. Closed circles are for $h\ k\ l$ that are all even or all odd. The result is that the reciprocal unit cell is body-centred.

The first term is the Fourier transform of the basis alone and this is multiplied by the second term which is the transform of the all-face-centred lattice. This second term

$$1 + e^{i\pi(h+k)} + e^{i\pi(h+l)} + e^{i\pi(k+l)} \neq 0 \tag{3.82}$$

provided that h, k and l are all even or all odd integers (h, k and l must be all permutable).

You will see from this that this condition is going to be true for *any* all-face-centred crystal and independently of *any* crystal system. It is a general condition that can be used therefore to identify the lattice type of any crystal, simply by searching for those reflections that appear to be systematically absent. Generalizing this to all lattice types we obtain table 3.3.

Note that the reflection condition for any all-face-centred F lattice means that reflections such as 100, 010, 001 etc will be absent in the diffraction pattern (these points in reciprocal space will be missing). Figure 3.38 shows what happens to the reciprocal lattice when this condition applies. The black points are the reciprocal lattice points for which the indices are either all odd or all even, whereas the open circles represent those points that are not allowed. It can therefore be seen that the unit cell in reciprocal space has become body-centred. In other words, the Fourier transform of an all-face-centred lattice is a body-centred lattice and vice versa. Therefore, we have the following relationships between real and reciprocal lattices.

P	↔	P
A	↔	A
B	↔	B
C	↔	C
I	↔	F
F	↔	I

Returning now to equation (3.81), let us take a closer look at the first term, the basis transform, and ask the question whether this too can become equal to zero. In other words, we wish to determine the conditions when

$$f_C\left[1 + e^{i\pi\left(\frac{h+k+l}{2}\right)}\right] = 0 \qquad (3.83)$$

Multiplying this term by its complex conjugate to calculate the intensity contribution

$$\left[1 + e^{i\pi\left(\frac{h+k+l}{2}\right)}\right]\left[1 + e^{-i\pi\left(\frac{h+k+l}{2}\right)}\right] = 2 + 2\cos\pi\left(\frac{h+k+l}{2}\right) = 0 \qquad (3.84)$$

whence

$$\cos\pi\left(\frac{h+k+l}{2}\right) = -1$$
$$\therefore \frac{h+k+l}{2} = 2n+1 \qquad (3.85)$$
$$\text{or } h+k+l = 4n+2$$

Thus, reflections like 222 will be systematically absent in the diffraction pattern. This is not a lattice condition but a special condition caused by accidental cancellation of terms in the molecular transform, and so is specific to the diamond type of structure. If we change the two carbon atoms in the molecule so that they are different, for example In and As, then the 222 reflection would be allowed, but the all-face-centering conditions would still apply.

By the way, the fact that the 222 reflection is missing for this structure is used in the design of crystal monochromators. For example, a Ge crystal cut on the (111) plane is often used because a white beam of radiation incident on it will in general pass intensity corresponding to the wavelengths λ, $\lambda/2$, $\lambda/3$, etc. Because the 222 intensity is equal to zero, this means that the harmonic intensity due to $\lambda/2$ will not be present, and the higher harmonics will be sufficiently weak so as to not contribute too much (in fact, the absence of 222 intensity is not strictly true. A very weak 222 signal can be seen by x-ray diffraction. Remember our assumption that the atomic electron density is spherical? But see later!).

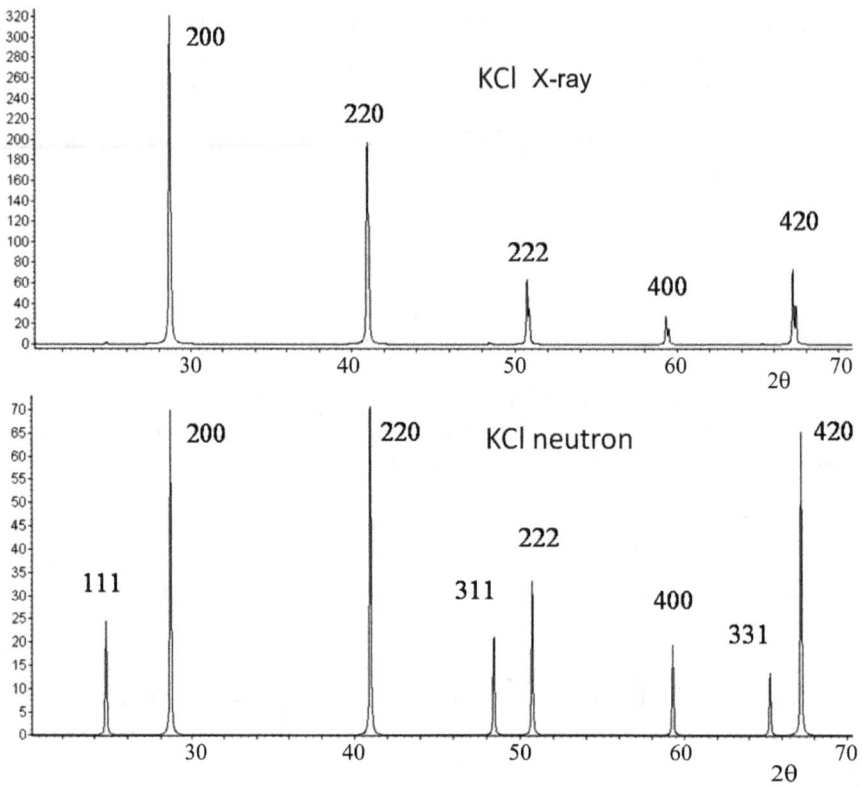

Figure 3.39. Computed powder diffraction patterns for KCl using x-rays and neutrons. $\lambda = 1.5418$ Å. The splitting seen for peaks in the x-ray pattern, especially as 2θ increases, arises from the fact that the x-ray beam is a close doublet CuKα1 and CuKα2, with $\lambda = 1.540\,51$ Å and $1.544\,03$ Å, respectively

Table 3.4. Indexing of KCl x-ray powder pattern.

2θ	d (Å)	$1/d^2$	Integer Ratio	$h\,k\,l$
28.642	3.114	0.1031	1	100
40.951	2.202	0.2062	2	110
50.74	1.798	0.3094	3	111
59.30	1.557	0.4125	4	200
67.16	1.393	0.5156	5	210

Consider, now, the difference in x-ray and neutron powder diffraction patterns for KCl, which we know to have an all-face-centred cubic lattice. This is shown in figure 3.39. Note that the neutron pattern appears to show many more peaks than the x-ray pattern. Table 3.4 shows the measured 2θ values for the x-ray pattern. In order to index the lines, we first of all convert the 2θ values to interplanar spacings d using Bragg's Law and then construct the set of $1/d^2$. As we know that KCl is cubic we can then find the ratios of these values (rounded to the nearest integer) and then the indices from

$$h^2 + k^2 + l^2 = \frac{a^2}{d^2} = \text{Ratio} \qquad (3.86)$$

This list of indices suggests that KCl is primitive cubic with $a = 3.114$ Å. However, if we repeat the process with neutron data, the sequence of $h\,k\,l$ values is now 111, 200, 220, 311, 222, 400, 331 and 420. These indices are all even or odd, indicating that the lattice is all-face-centred and that $a = 6.228$ Å. Therefore, the indices used in the x-ray pattern need to be doubled in order to be consistent with the neutron data. The cause of this difference is that x-rays are unable to distinguish easily between scattering by K^+ and Cl^- ions, since they each contain the same numbers of electrons and hence have very similar x-ray atomic scattering factors. Had the diffraction pattern been from, say, NaCl, which is of the same structure type, this ambiguity would not have occurred.

3.14 Solution of crystal structures

I miss the old days, when nearly every problem in x-ray crystallography was a puzzle that could be solved only by much thinking.

Linus Pauling

The phase problem

In order to determine crystal structures (see [11–13]), i.e., the location of all the atomic positions in a crystal, we need to calculate the density $\rho(xyz)$ at any place in the unit cell from the known structure amplitudes using the standard Fourier transformation formula:

$$\rho(\mathbf{r}) = \int F(\mathbf{Q}) e^{-2\pi i \mathbf{r} \cdot \mathbf{Q}} \, d\mathbf{Q} \qquad (3.87)$$

Notice the minus sign in the exponent: this is because we are now changing from reciprocal space back to real space. Thus, corresponding to the structure factor formula (3.69), we can write

$$\rho(xyz) = \frac{1}{V} \sum_{\overline{hkl}}^{hkl} F(hkl) e^{-2\pi i (hx+ky+lz)} \qquad (3.88)$$

V is the volume of the unit cell. Note that the summation is over all reflections from $\overline{hkl}$ to hkl, i.e., over all possible reflections.

In x-ray diffraction, $\rho(xyz)$ is the electron density at the point (x, y, z) in the unit cell. In neutron diffraction, it signifies the nuclear density. However, this formula is not particularly suitable for calculating the electron density in practice because it contains complex quantities. Consider the term in the summation for the moment

$$F(hkl) e^{-2\pi i (hx+ky+lz)} = e^{-2\pi i (hx+ky+lz)} \sum_j f_j e^{2\pi i (hx_j+ky_j+lz_j)} \qquad (3.89)$$

which can be written as

$$\{\cos 2\pi(hx + ky + lz) - i \sin 2\pi(hx + ky + lz)\}\{A(hkl) + iB(hkl)\} \quad (3.90)$$

where

$$A(hkl) = \sum_j f_j \cos 2\pi(hx_j + ky_j + lz_j)$$
$$B(hkl) = \sum_j f_j \sin 2\pi(hx_j + ky_j + lz_j) \quad (3.91)$$

Combining with the complex conjugate we get

$$\{A(hkl) + iB(hkl)\}\{\cos 2\pi(hx + ky + lz) - i \sin 2\pi(hx + ky + lz)\}$$
$$+ \{A(\bar{h}\bar{k}\bar{l}) + iB(\bar{h}\bar{k}\bar{l})\}\{\cos 2\pi(hx + ky + lz) + i \sin 2\pi(hx + ky + lz)\} \quad (3.92)$$
$$= 2A(hkl)\cos 2\pi(hx + ky + lz) + 2B(h\,k\,l)\sin 2\pi(hx + ky + lz)$$

since $A(\bar{h}\bar{k}\bar{l}) = A(hkl)$ and $B(\bar{h}\bar{k}\bar{l}) = -B(hkl)$. Then

$$\rho(x\,y\,z) = \frac{1}{V}\left[2\sum_{h\bar{k}\bar{l}}^{hkl}[A(h\,k\,l)\cos 2\pi(hx + ky + lz) + B(h\,k\,l)\sin 2\pi(hx + ky + lz)]\right] \quad (3.93)$$

The summation here is taken only for reflections with $h > 0$, i.e., over half of reciprocal space. Note that $F(000)$ is excluded from this formula. We now introduce the phase angle $\phi(hkl)$ defined by

$$\tan \phi(hkl) = \frac{B(hkl)}{A(hkl)} \quad (3.94)$$

and then

$$\rho(xyz) = \frac{2}{V}\sum_{h\bar{k}\bar{l}}^{hkl}|F(hkl)| \cos\{2\pi(hx + ky + lz) - \phi(hkl)\} \quad (3.95)$$

This result shows that to compute the density $\rho(x\,y\,z)$ it is necessary to know the values of the structure amplitudes |**F**(hkl)| and also the phases $\phi(hkl)$. The structure amplitudes come directly from the measurements of the relative intensities

$$I(hkl) \propto |F(hkl)|^2 \quad (3.96)$$

but in general the phases are not known. This *phase problem* has been a challenge towards automatic structure determination over the last century. Some of the methods developed to handle this are summarized below.

1. Trial and error—guess a model or part model and then to reveal the remaining atoms construct a Fourier map based on the *difference* between observed and calculated structure amplitudes to find any remaining atoms

$$\Delta\rho(xyz) = \frac{2}{V}\sum_{h\bar{k}\bar{l}}^{hkl}[|F_{obs}(hkl)| - |F_{cal}(hkl)|] \times \cos\{2\pi(hx + ky + lz) - \phi(hkl)\} \quad (3.97)$$

2. Heavy atom method—if there is a heavy atom, e.g. Hg, present in the structure the x-ray intensities are so dominated by its contribution that one can consider the phases to be essentially due to this single atom alone. So, the calculated phases based on a crystal structure consisting of this atom alone can be used together with the observed amplitudes in order to obtain an initial model.
3. Isomorphous replacement—compare diffracted intensities from two crystals in which one atom type has been replaced by another. This is particularly useful in protein structure determination. This replacement can be done by chemical means or by tuning the wavelength of incident synchrotron radiation.
4. Anomalous dispersion—makes use of the fact that atomic scattering factor actually contains real and imaginary parts that are large especially close to an x-ray atomic absorption edge. This has the effect of breaking Friedel's Law.
5. Direct methods—uses the statistical distribution of intensities in reciprocal space to find relationships between phases of different reflections. Thus, if one starts with a couple of trial phases, direct methods enable likely phases of other reflections to be found. This is particularly useful for small molecule problems and is widely used as a nearly automatic method of solving crystal structures. The pioneering work on this resulted in the Nobel Prize being awarded in 1986 to Jerome Karle and Herman Hauptmann.

We shall see below a couple of other ways in which the phase problem has been handled, but before this let's consider how the formula (3.95) can be used to locate the atomic positions, assuming that the phases are known.

3.15 Fourier synthesis

To obtain a map of the density representing the contents of the unit cell in a crystal structure it is necessary to add together all of the waves described by amplitudes $|F(h\ k\ l)|$ together with their phases $\phi(hkl)$. As a simple example, consider figures 3.40(a)–(d) where we sketch the effect of adding four plane waves together all with phase angle $\phi = 0°$.

In (a) the plane wave for the 100 reflection is sketched. In this case the peaks lie on the (100) planes with the troughs halfway between. The amplitude $|F(hkl)|$ is a measure of the height of this wave, represented by the amount of shading. In (b) the plane wave for the 010 reflection has been added, in (c) the wave corresponding to the 110 reflection, and finally in (d) the wave for the 1̄10 reflection. It can be seen that with these four reflections alone it is already apparent that density is being built up on the corners of the unit cell with a weaker component at the centre of the unit cell. In (e) the phases for the 110 and 1̄10 reflections have been changed by $\pi/2$. By shifting the maxima of these two waves density is now seen to form halfway along the unit cell axes, with little density at the centre of the unit cell, a completely different structure from before.

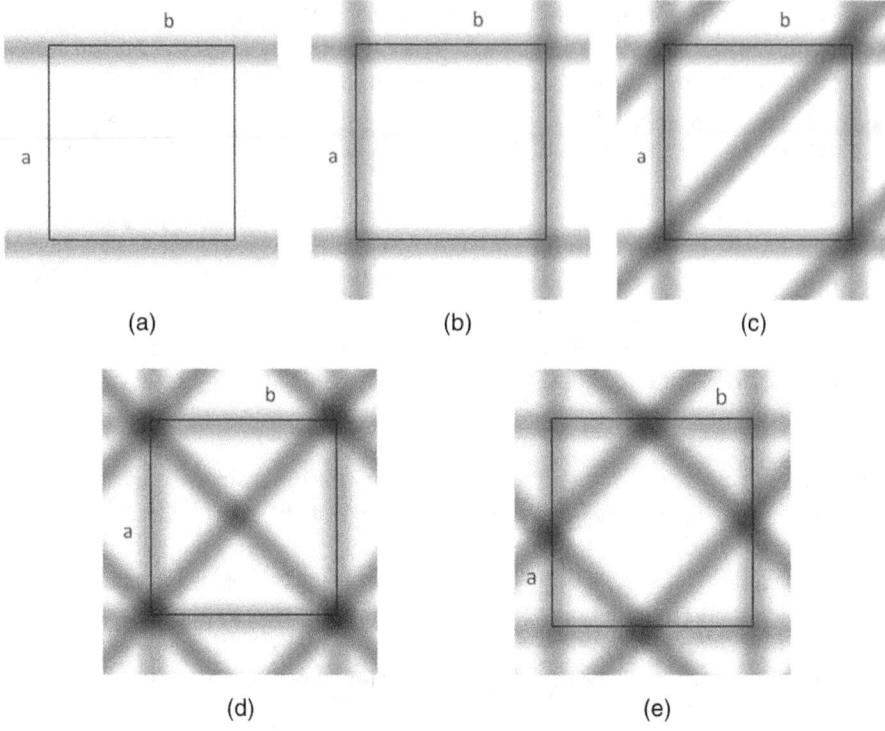

Figure 3.40. Example of addition of planes waves in Fourier synthesis of electron density.

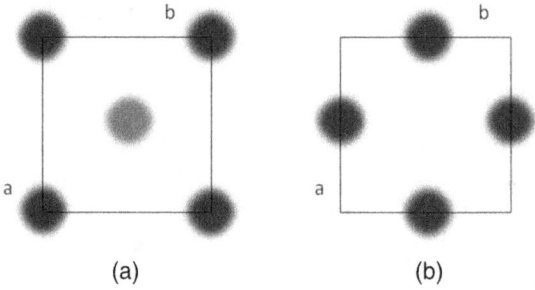

Figure 3.41. Fourier synthesis with infinite number of reflections.

In figure 3.41 is sketched what might be the result of summing the Fourier series over a very large number of *hkl* reflections starting from figure 3.40(d) and (e). The effect is to concentrate the density into the atomic positions with little or no observable density between. We end up with something resembling a photograph of the atoms in the structure. The main difference obtained between the two structures shows that it is the phases that are crucial for correct structure determination, rather than simply the structure amplitudes.

Figure 3.42 illustrates the important fact that it is the phases that dominate in Fourier synthesis rather than the amplitudes of the waves. At the top, we see photographs of the Nobel Prizewinners Jerome Karle on the left, and Herbert

Figure 3.42. The relative importance of phases over amplitudes [14]. Reproduced from [15] with permission of Elsevier.

Hauptman on the right. Below this, the images have been reconstructed by swapping the phases but keeping the amplitudes. The result clearly shows that most of the information has been swapped too.

Before the advent of computers, various clever photographic techniques were invented to perform Fourier syntheses in this way. The author is sufficiently advanced in years to have used one called the 'von Eller photosommateur' during his graduate studies (figure 3.43). I present it here for its pedagogical value.

This consisted of a cylindrical light-tight can. At one end, a lamp illuminated a glass mask on which were a set of cosinusoidal fringes. The light then passed through this onto a photographic plate outside the can to which was taped a scale drawing of the reciprocal lattice. By rotating the photographic plate (and reciprocal lattice) and at the same time winding a cursor up and down it was possible to position the cursor directly onto a reciprocal lattice point. This had the effect of orienting the relevant reciprocal lattice vector perpendicular to the image of the cosine mask. At the same time, the mask moved forwards or backwards thus changing the apparent fringe spacing projected onto the photographic plate. Furthermore, left and right movements of the mask enabled the correct phase to be set up. The light was then exposed

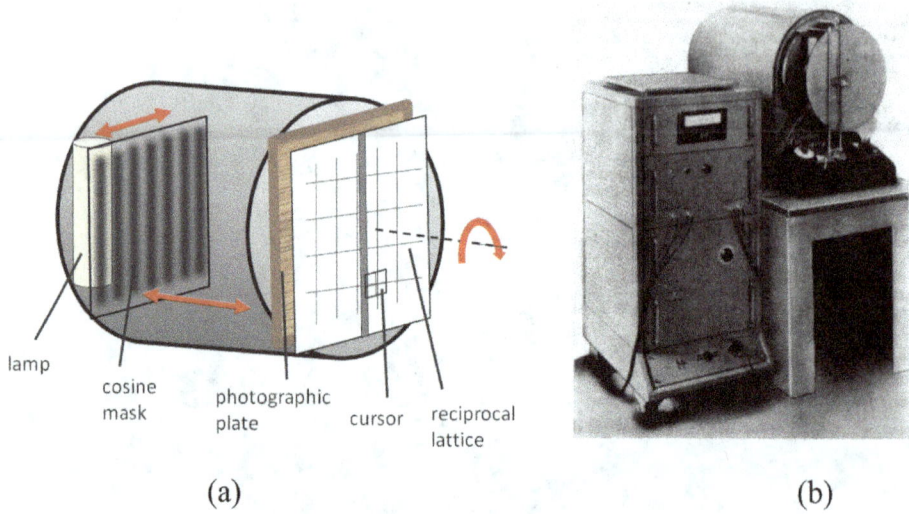

Figure 3.43. (a) Schematic diagram of von Eller photosommateur. (b) Photograph of the original photosommateur (reproduced with permission of the International Union of Crystallography).

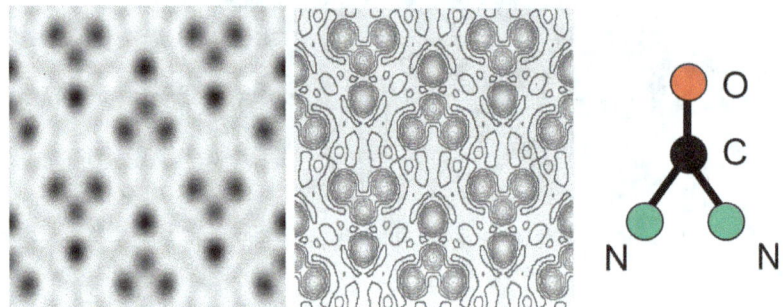

Figure 3.44. Fourier maps of the crystal structure of urea on (010): left performed with the von Eller photosommateur, middle computer generated [16], right the urea molecule (hydrogens omitted).

for several seconds proportionally to the structure amplitude for the relevant reflection, and the whole process repeated for the next reciprocal lattice point. The result was effectively a photograph of the atoms in the crystal structure. Figure 3.44 shows the result of using this technique to plot a projection on (010) of the crystal structure of urea together, for comparison, with a modern computer-generated contour map. Although one would not dream of using such a method today, it was nevertheless a wonderfully fun way to learn about Fourier syntheses [14].

Before the general use of computers became standard, prior to the mid-1960s, many different and cunning techniques like this were devised to locate the atomic positions from the data, even though the intensities of the reflections were often simply estimated by eye on an arbitrary scale! Today, with modern computers and new methods of collecting data, it is commonplace to determine crystal structures routinely with literally hundreds or even thousands of atoms in the unit cell. To do this of course, one has first to deal with the phase problem. Many techniques for doing this have been developed over the last decades. In the next sections, we shall look at just a couple of methods by which this has been done successfully.

3.16 The Patterson method

This method of getting at the phases was first demonstrated by Arthur Lindo Patterson in 1934 in the USA. Consider for the moment a one-dimensional electron density at two positions $\rho(x)$ and $\rho(x + u)$. The average product (or autocorrelation) in a repeat of length a is then given by the formula

$$A(u) = \int_0^1 \rho(x)\rho(x + u)\mathrm{d}x \tag{3.98}$$

Therefore

$$\begin{aligned} A(u) &= \int_0^1 \frac{1}{a^2} \sum_h F(h)e^{-2\pi i h x} \sum_{h'} F(h')e^{-2\pi i h'(x+u)} \mathrm{d}x \\ &= \frac{1}{a^2} \sum_h F(h) \sum_{h'} F(h')e^{-2\pi i h' u} \int_0^1 e^{-2\pi i (h+h')x} \mathrm{d}x \end{aligned} \tag{3.99}$$

But since h and h' are integers

$$\int_0^1 e^{-2\pi i(h+h')x} \mathrm{d}x = \left[\frac{e^{-2\pi i(h+h')x}}{-2\pi i(h + h')} \right]_0^1 = 0 \tag{3.100}$$

except for when $h = -h'$. Therefore

$$A(u) = \frac{1}{a^2} \sum_h \sum_{-h} F(h)F(-h)e^{2\pi i h u} \tag{3.101}$$

Now, we also know that from Friedel's Law

$$F(-h) = F^*(h) \tag{3.102}$$

and so

$$A(u) = \frac{1}{a^2} \sum_h |F(h)|^2 \, e^{2\pi i h u} \tag{3.103}$$

This can be rewritten as

$$A(u) = \frac{1}{a^2} \sum_{h>0} (|\mathbf{F}(h)|^2 \, e^{2\pi i h u} + |\mathbf{F}(h)|^2 \, e^{-2\pi i h u}) \tag{3.104}$$

and using Friedel's Law

$$A(u) = \frac{2}{a^2} \sum_{h>0} |F(h)|^2 \cos 2\pi h u \qquad (3.105)$$

For convenience, we now define the Patterson function $P(u)$ by

$$P(u) = \frac{2}{a} \sum_{h>0} |F(h)|^2 \cos 2\pi h u \qquad (3.106)$$

In three-dimensions this becomes

$$P(uvw) = \frac{2}{V} \sum_{h>0} \sum_{k} \sum_{l} |F(hkl)|^2 \cos 2\pi(hu + kv + lw) \qquad (3.107)$$

The important thing about this result is that no phase information is involved. Because of the starting point as an autocorrelation between two electron densities at different positions, the Patterson function provides information on *vectors* between atoms, but not on individual atoms themselves. Thus, one can construct a Fourier map using this function, rather than the function for electron density, by using the squares of structure amplitudes with all plane waves having their phases set to zero. It is obvious that in such a map all waves will add together at the origin of the Patterson unit cell, corresponding to self-vectors (i.e. vectors of zero length corresponding to the atomic distance from each atom to itself). Peaks elsewhere in the map will correspond to vectors between different atoms. I find that a useful way to think of this is to remember the following quotation I heard whilst attending lectures at Birkbeck College London by the crystallographer C H (Harry) Carlisle: 'All vectors to a common origin'. He used to call this the 'peasant's definition of the Patterson function'. Let's see an example in practice to get the idea.

Figure 3.45 shows schematically how to build up a Patterson map from a known structure. On the left-hand side is shown a hypothetical crystal structure consisting of three atoms, coloured red, yellow and green, with decreasing numbers Z_1, Z_2, Z_3 of electrons, respectively. The right-hand side shows the Patterson map in the process of construction. Starting with the red atoms alone, the self-vector (red atom to red atom) creates a peak at the origin of the Patterson cell approximately of height Z_1^2. Now consider the vector from a red atom at the corner of the unit cell to the nearby yellow atom (top diagram). Placing the origin of this vector at the Patterson cell origin creates a peak in the Patterson map of height Z_1Z_2 (remember 'all vectors to a common origin'). At the origin, we also add Z_2^2 for the yellow atomic self-vector However, one has also to consider the reverse vector from yellow atom to red atom and this creates another peak of height Z_1Z_2 outside the Patterson cell, which by translational symmetry gives a peak inside the lower right of the cell (middle diagram). Finally, we add the vectors green to red, red to green, green to yellow and yellow to green to lead to the distribution of Patterson peaks shown at bottom right of the figure. The result, despite having no knowledge of the phases of the reflections, gives

A Journey into Reciprocal Space

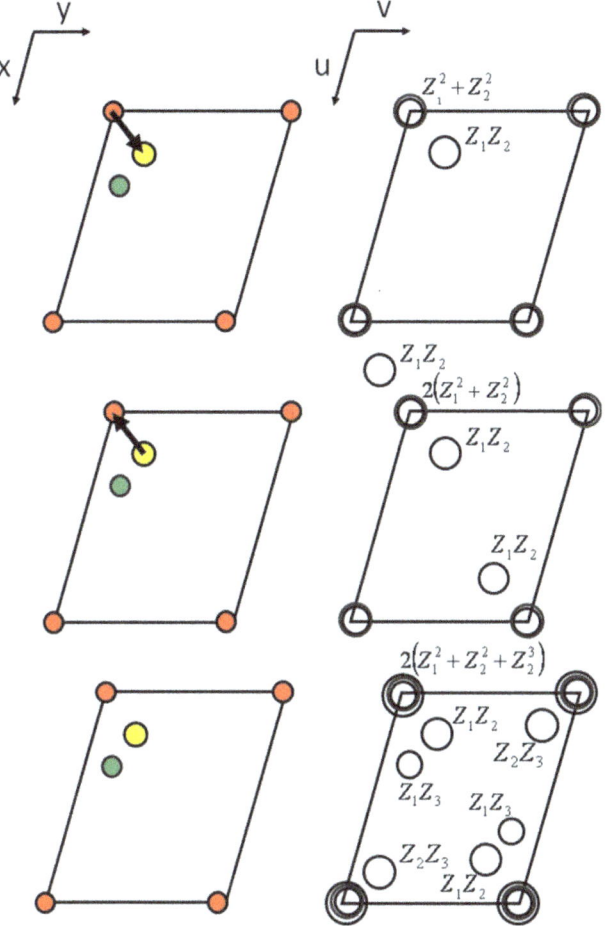

Figure 3.45. Construction of a Patterson map.

rise to a map on which the peaks are closely related to the underlying crystal structure. It is then the task of the crystallographer to deconvolute this map to reveal the most likely model for the structure. This is not always unique, since different models can give very similar Patterson maps, but at least the number of models is finite. These days, solving Patterson maps can be done automatically by computer programs. Note, by the way, that the Patterson map is always centrosymmetric because for each vector in one direction there will always be another the opposite way.

Figure 3.46 shows a simulation using a Java applet that can be downloaded from [17]. You may care to try this out yourself in order to familiarize yourself with the Patterson function.

3.17 Charge flipping

This is one of the most recent and possibly most exciting developments in phase determination, and was demonstrated by two Hungarian optical physicists Oszlányi

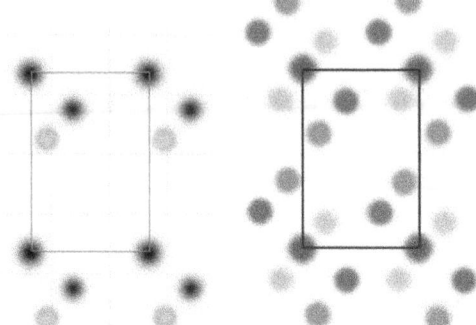

Figure 3.46. Simulation of Patterson map from a crystal structure. Left real structure, right Patterson map.

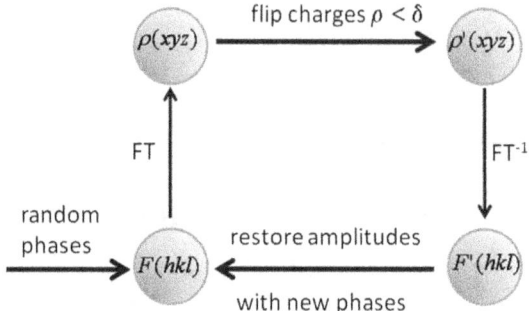

Figure 3.47. Algorithm for charge flipping.

and Sütő in 2004 [18]. The basic idea behind this owes much to algorithms developed years earlier for optical image processing. The procedure is amazingly simple to follow in principle and to code into software. The charge flipping algorithm is an iterative process starting with a complete set of diffraction intensities. The algorithm does not require any *a priori* information about the symmetry of the structure or its chemical composition. These properties can be included in the model later after the application of charge flipping and can be, even to a large extent, derived directly from the result of charge flipping. The iterative process is illustrated in figure 3.47.

First of all, start with the set of observed structure amplitudes |F(hkl)| and assign random phases to each. Fourier transformation is then used to create an electron density map $\rho(xyz)$ which initially will appear to be unrecognizable. A cut-off density δ is assigned and all densities below this value have their signs reversed to produce a new map $\rho'(xyz)$. Fourier transformation of this gives a new set of structure amplitudes |F'(hkl)| with new phases. We now take the new phases and apply them to the original set of observed structure amplitudes. The cycle is then repeated continuously until density maps showing the crystal structure are obtained. The whole process seems to work like magic and is very entertaining to watch.

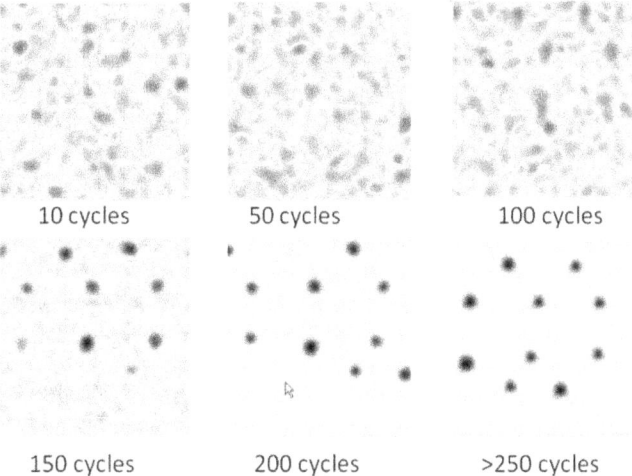

Figure 3.48. Sequence of electron density calculations as function of number of cycles. In the last picture, the origin has been shifted for convenience to show the whole molecule.

Figure 3.48 shows a few slides taken from a Java applet available at [17]. The reader is strongly encouraged to try this out.

3.18 The Rietveld method

In 1969 Hugo Rietveld [19] published a landmark paper in the field of powder diffraction analysis. Up to this time, powder diffraction had principally been used as a means of identifying materials and obtaining lattice parameter information, and generally held little interest beyond this. However, Rietveld showed that if one had a reasonable model of the crystal structure of a material, it was possible to refine by least-squares the positions of the atoms by fitting a calculated powder pattern to the observed data without the need to physically separate the individual structure factors making up the peaks. This launched an enormous renewed interest in powder diffraction so that these days there are whole conferences devoted to this subject. The Rietveld method is nowadays a fully accepted and standard method to refine crystal structure information. It is particularly useful for situations where single crystals are not readily available (by far the majority of cases). The basic idea is to consider the contribution to the intensity y_i of the measured peak profile at positions $2\theta_i$. Assuming a Gaussian shape to the profiles

$$y_i = tF^2(hkl)m_j L_j \frac{2\sqrt{\ln 2}}{H_j \sqrt{\pi}} e^{-4\ln 2\{(2\theta_i - 2\theta_j)/H_j\}^2} \qquad (3.108)$$

where
    t = step width of the counter
    m_j = multiplicity of jth reflection
    L_j = Lorentz factor (for x-rays one needs to include a polarization factor)

$2\theta_j$ = calculated positions of the jth Bragg peak
H_j = full width of peak at half-height.

This Rietveld equation has since been modified for different peak shape functions and other parameters have been added, but the basic idea remains true. The parameter H_j is often given by the relationship

$$H_j^2 = U \tan^2 \theta_j + V \tan \theta_j + W \qquad (3.109)$$

where U, V and W are parameters that can be refined during the process. We see from this that the parameters to be refined can be considered in two parts

1. $F(hkl)$ which contains all the structural parameters of interest (coordinates, displacement parameters, occupation factors etc). Also $2\theta_j$ is determined by refining the unit cell parameters.
2. Geometrical factors to fit the profile pattern including any zero error, H_j (with parameters U, V and W), absorption correction parameters, preferred orientation, peak asymmetry etc.

For most users of the method only the parameters in (1) are of any real interest, while the geometrical factors (2) are often ignored as they are mainly parameters that are needed to help with the fitting of the profiles.

Figure 3.49 shows a typical Rietveld fit, in this case to a neutron powder diffraction pattern from a mixture of three different substances. The Rietveld analysis not only gives a good fit to the data but can also estimate the percentage of each component. In addition to using the Rietveld process on angle-dispersive data such as this, it has also been applied to neutron time-of-flight data (figure 3.50).

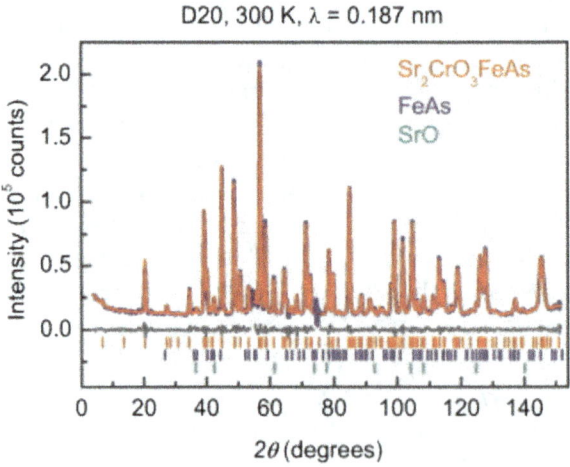

Figure 3.49. Rietveld plot for a neutron powder pattern of a mixture of compounds). Red marks the fitted data; blue is measured data. A plot of the difference between observed and calculated patterns is shown immediately below, and below this the positions of the reflections for the three components in the mixture are marked. Reproduced from [20] with permission.

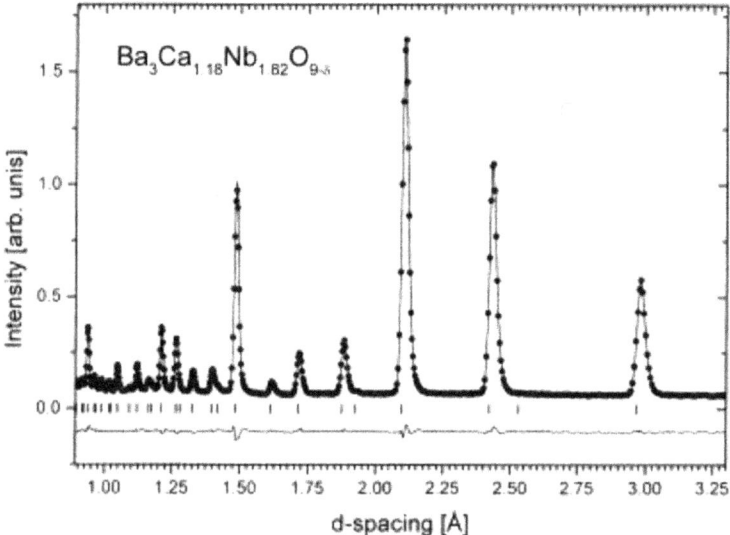

Figure 3.50. Neutron time-of-flight Rietveld fit for $Ba_3Ca_{1.18}Nb_{1.82}O_{9-\delta}$. Reproduced from [21] with permission of Elsevier.

3.19 Total scattering analysis

A more recent development in the way crystal structure information can be extracted from powder diffraction is the so-called total-scattering or PDF (pair distribution function) analysis. Essentially this is to consider the whole diffraction pattern, peaks plus background, and treat it as if one were dealing with diffraction from a liquid or amorphous material. The method has usually been applied with neutron scattering, although x-rays can also be used. For this method to work, it is necessary for the data to extend to very high values of **Q**. While this can be achieved easily with time-of-flight neutron sources, it is more difficult with conventional x-ray sources.

Suppose there are n different atomic species in a material. Let c_j be the concentration of atomic species j and b_j the effective coherent bound neutron scattering length. The total scattering factor $F(\mathbf{Q})$ is related to the total radial distribution $g(\mathbf{r})$ by

$$F(\mathbf{Q}) = \langle \rho \rangle \int_0^\infty 4\pi r^2 G(\mathbf{r}) \frac{\sin Qr}{Qr} dr \qquad (3.110)$$

where $\langle \rho \rangle$ is the average number density (in atoms $Å^{-3}$).

Definition: *The radial distribution function, $G(\mathbf{r})$, or pair correlation function, is a measure of the correlation between atoms within a system. It is a measure of the average probability of finding an atom at a distance r away from a given reference atom. It determines how many atoms are within distances of r and r + dr away from an atom.*

$G(\mathbf{r})$ can be defined in terms of partial radial distribution functions $G_{ij}(\mathbf{r})$:

$$G(\mathbf{r}) = \sum_{i,j=1}^{n} c_i c_j \langle b_i \rangle \langle b_j \rangle \left[G_{ij}(\mathbf{r}) - 1 \right] \qquad (3.111)$$

where

$$G_{ij}(\mathbf{r}) = \frac{n_{ij}(\mathbf{r})}{4\pi r^2 \rho_j d\mathbf{r}} \qquad (3.112)$$

$n_{ij}(\mathbf{r})$ is the number of atoms of type j within distances of r and $r + dr$ of an atom of type i, and $\rho_j = c_j \langle \rho \rangle$. For an ordered crystal, the function $G(\mathbf{r})$ is a series of sharp peaks with relatively low background. However, as disorder is introduced into the structure the peaks become broader. Nonetheless their positions as a function of distance r provide useful information about atomic separations.

In figure 3.51 are shown some total scattering results for the important piezo-electric ceramic $PbZr_{1-x}Ti_xO_3$ (PZT). In (a) the experimental $G(r)$ is shown as a function of distance r. The red curve is a fit to the observed data using a special program called PDFFIT using the known average crystal structure, assumed to be rhombohedral in nature, as derived from Rietveld refinement. However, the structure of PZT has always been controversial, with some evidence that it is in fact monoclinic. (b) shows the fit for the monoclinic structure. (c) shows the result of what is known as reverse Monte Carlo (RMC) modelling. Here a computer simulation is set up in which there is a large array of unit cells containing the atoms in their average positions. The atoms are then allowed to move within certain limits while monitoring a fit to the total scattering and the peaks seen in the Rietveld analysis. It can be seen that the fit is very good. In (d) two peaks in $G(r)$ have been identified as corresponding to the Zr–O and Ti–O distances, and in (e) the O–O and Pb–O distances are observed. (f) shows the PDFFIT to the first few peaks where it can be seen that there is considerable disagreement with the experimental plot. However, once one allows for the Pb positions to have a degree of disorder in the structure, an excellent fit is obtained. The results of this analysis have played an important role in understanding the property of enhanced piezoelectricity in ceramic PZT.

It can be seen that total scattering provides a large amount of useful information about crystal structures on the local scale, which, when taken together with the long-range average structure, can be of great use in understanding the role of atomic positions and displacements in the physical properties of materials.

3.20 Aperiodic crystals

I told everyone who was ready to listen that I had material with pentagonal symmetry. People just laughed at me.

Dan Shechtman

Just when you think that everything about crystal symmetry must be known, nature retaliates by throwing a curve ball. We have already seen that the basic idea of a periodic lattice is a property defining real crystals, but the scientific world was

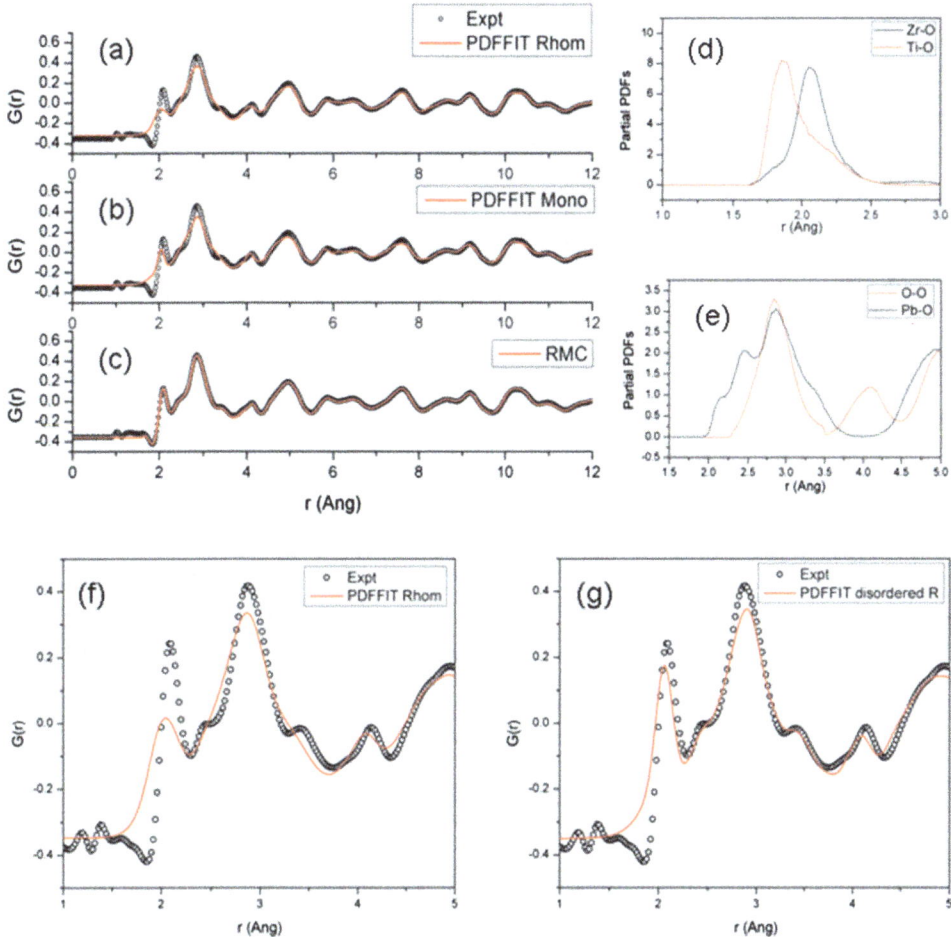

Figure 3.51. (a) PDFFIT for $x=0.3$ at room temperature using a rhombohedral structure, (b) using a monoclinic structure. (c) Fitting by reverse Monte Carlo method. (d) Partial PDF peaks for Zr–O and Ti–O. (e) Partial PDF peaks for Pb–O and O–O from the RMC modelling. (f) Enlargement of (a) showing the disagreement between the observed and fitted results. (g) Addition of disordered Pb displacements around the [111] axis now gives rise to the shoulders on the main Pb–O peak (from [22]).

unprepared for a major discovery which subsequently altered our perception of what is meant by a crystal.

In 1984, a rapidly cooled metallic Al_4Mn alloy was observed by the Israeli-born materials scientist, Dan Shechtman (1941–), to exhibit 10-fold symmetry in its electron diffraction pattern. This seemed to show that these materials in some way violated the rule that crystal lattices could not show 5- or 10-fold symmetry. Such materials were termed *quasiperiodic crystals* or *quasicrystals*, and they clearly could not be explained by conventional crystal symmetry ideas. Originally, this discovery was met with disbelief, and Shechtman's original paper was rejected for publication; even the Nobel laureate Linus Pauling claimed that the effect could be explained as a form of multiple twinning, a mixture of crystals in different orientations. However,

Shechtman's work was extremely precise and methodical, and to his credit he pursued this discovery against all the opposition. He was subsequently proved correct and was eventually awarded the 2011 Nobel Prize in Chemistry for this discovery. We now know that such quasicrystals are not particularly rare, especially in metal alloys, and one can even grow crystals with observable faces arranged in 5-fold symmetries. For example, the alloy $Al_{63}Cu_{24}Fe_{13}$ can be grown as a faceted, single quasicrystal up to 1 mm^3 in size.

How then can we explain this apparent violation of basic crystal lattice symmetry? A convenient way to do this is to think of different ways in which to tile a floor without leaving gaps. In the conventional approach, if we use a set of identical regular tiles, one ends up with periodicity and the restriction that 5- and 7-fold symmetries are not allowed. However, if one is prepared to use differently shaped tiles, then one can obtain many arrangements with local 5- or 7-fold symmetries. Such arrangements are not periodic in the traditional sense, but nonetheless create arrangements that are ordered in that they occur according to certain rules. This has an old history in fact. As far back as 1619, Kepler showed how to fill a two-dimensional space with *different* 5-fold symmetric tiles.

Figure 3.52 shows an example of tiling in two dimensions where two different shapes are used, from an idea by Roger Penrose (1931–). We see that with appropriate rules for stacking, fat and thin rhomb shapes can be placed together to fill two-dimensional space. Such an arrangement never repeats, although it does show local orientational symmetry of a 5-fold kind.

Interestingly, if one superimposes a series of lines, due to Robert Ammann (1946–94), as shown, we find that the sets of lines together show 5-fold symmetry! Furthermore, the lines are spaced by long (L) and short (S) distances in the sequence

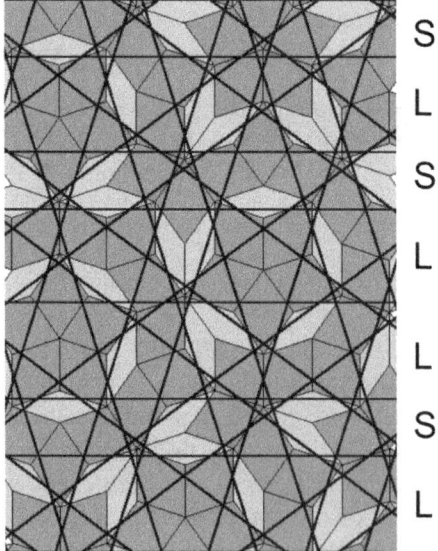

Figure 3.52. Penrose tiling of two types of rhombic cell. Reproduced from [23] with permission of Oxford University Press.

L S L L S L S L… This sequence has an interesting property. If you replace every L by L S and every S by L, you find the sequence L S L L S L S L L S L L S….This is simply a copy of the original sequence. This type of sequence is found in many places in nature and was discovered by the Italian mathematician Leonardo Pisano Bigollo (c.1170–c.1250), also called Leonardo of Pisa, or simply Fibonacci. It is interesting to note that the ratio L/S is the 'golden mean $(1+\sqrt{5})/2 = 1.618\,03\ldots$, well known to artists who use this ratio to design a pleasing layout for a painting. Thus, we see that this arrangement is not random, nor is it periodic. Instead, the term *quasiperiodic* is used to describe it.

Notation: *The term aperiodic applies to those cases where long-range periodicity is not a feature of the solid, such as in amorphous (non-crystalline) materials like glass. The term aperiodic crystal is used to describe crystals where translational periodicity does not apply but where, nonetheless the atoms or molecules are arranged in some sort of order governed by different rules from the usual periodic rules.*

The subject becomes even more intriguing, since quasiperiodicity can be shown to arise by considering a regular periodic lattice of points in a six-dimensional space. If a cut is then made through such a lattice and projected onto two- or three-dimensional space, obviously the projection will consist of points. If the cut and projection are made in the right way, it turns out that an array of points can be obtained with, say, local 5-fold symmetries, corresponding to the observed diffraction pattern. In addition, the positional ordering of the points along particular directions is again found to obey a rule such as for the Fibonacci sequence. It is important to understand that the atoms or molecules in aperiodic crystals are arranged, nonetheless, in an ordered fashion.

As a consequence of Shechtman's discovery, in 1992 the International Union of Crystallography [24] revised the definition of a crystal in terms of reciprocal space.

Definition: *A material is a crystal if it has **essentially** a sharp diffraction pattern. The word **essentially** means that most of the intensity of the diffraction is concentrated in relatively sharp **Bragg peaks**, besides the always present diffuse scattering. In all cases, the positions of the diffraction peaks can be expressed by*

$$\mathbf{H} = \sum_{i=1}^{n} h_i \mathbf{a}_i^* \quad (n \geqslant 3) \tag{3.113}$$

Here $\mathbf{a}_i^*$ and h_i are the basis vectors of the reciprocal lattice and integer coefficients, respectively, and the number n is the minimum for which the positions of the peaks can be described with integer coefficient h_i.

The conventional crystals are a special class, though very large, for which $n = 3$.

Normal crystals in three dimensions are when $n = 3$, whereas aperiodic crystals are described as periodic in a higher-dimensional space, $n > 3$, with symmetries described by appropriate higher-dimensional space groups. It is the projection into three-dimensional space from a higher dimension that creates the aperiodicity.

Aperiodic crystals also include the so-called incommensurate crystals, where superimposed on a fundamentally periodic arrangement there are long-range periodic wave-like disturbances to the atomic positions, so that the structure never actually repeats. This is shown in figure 3.53, where (a) shows a normal commensurate structure, here using the Statue of Liberty as a 'basis'. In (b) another commensurate structure is shown, but with twice the repeat determined by a wave-like disturbance. In (c) the disturbing wave is not commensurate with the fundamental repeats of the Statue of Liberty, thus creating a one-dimensional incommensurate structure. In diffraction space, this results in extra reflections that do not lie on the nodes of the usual reciprocal lattice, although here again, by going to a higher dimension the spots will lie on a regular lattice. For instance, consider the case of a three-dimensional reciprocal lattice with a one-dimensional additional modulation. From equation (3.113)

$$\mathbf{a}_4^* = \mathbf{q} = q_1\mathbf{a}_1^* + q_2\mathbf{a}_2^* + q_3\mathbf{a}_3^* \quad (3.114)$$

where at least one of the q_i is irrational to give extra reflections in the diffraction pattern. We see from this that such a reciprocal lattice with its additional irrational points can be treated as a true periodic lattice within a four-dimensional space. Such a crystal structure would then conform to a four-dimensional space group.

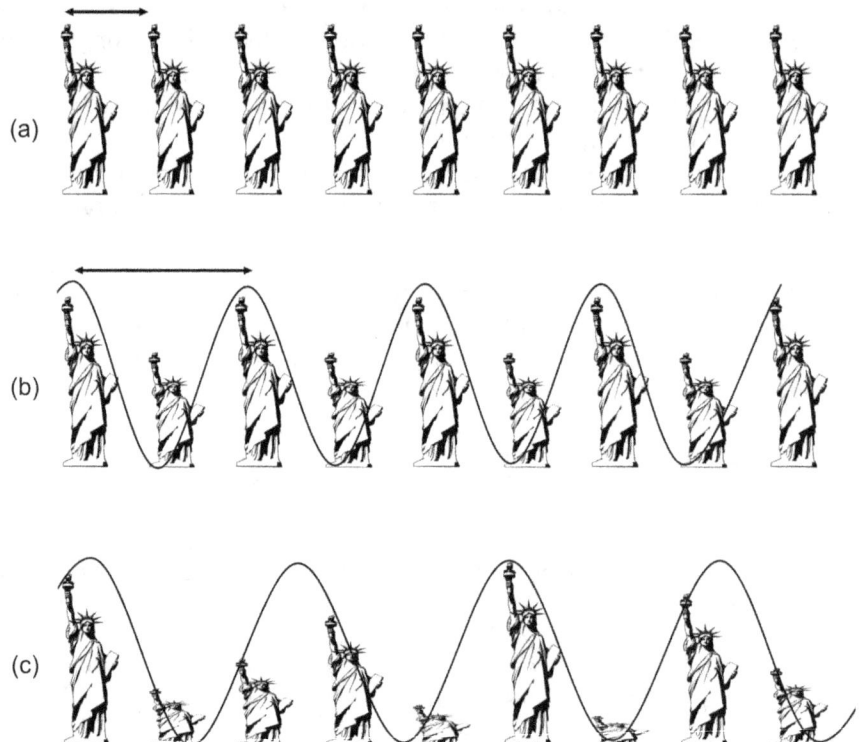

Figure 3.53. Commensurate and incommensurate modulations: (a) regular periodicity (b) commensurate modulation (c) incommensurate modulation.

3.21 Disordered crystals

So far, we have considered crystals in terms of the average positions of atoms in the solid. However, in reality the atoms are in a state of thermal vibration, and so the information contained within the Bragg intensities is essentially a time-average of the crystal structure. Furthermore, in many crystals there is spatial disorder, for example where certain atoms are replaced by different atoms, or certain molecules are differently oriented from the average. This disorder may not be completely random, but there may be some partial or short-range order, where statistically ordered arrangements exist in the crystal. The disordered part of the structure gives rise to extra, so-called diffuse, scattering superimposed on the Bragg intensity pattern, so that we can write the diffraction intensity as the sum of two terms:

$$I = I_{\text{Bragg}} + I_{\text{diffuse}} \qquad (3.115)$$

By conservation of energy, the more disorder that is present the lower I_{Bragg} and the higher I_{diffuse}. So, measurements of the Bragg intensities alone give information

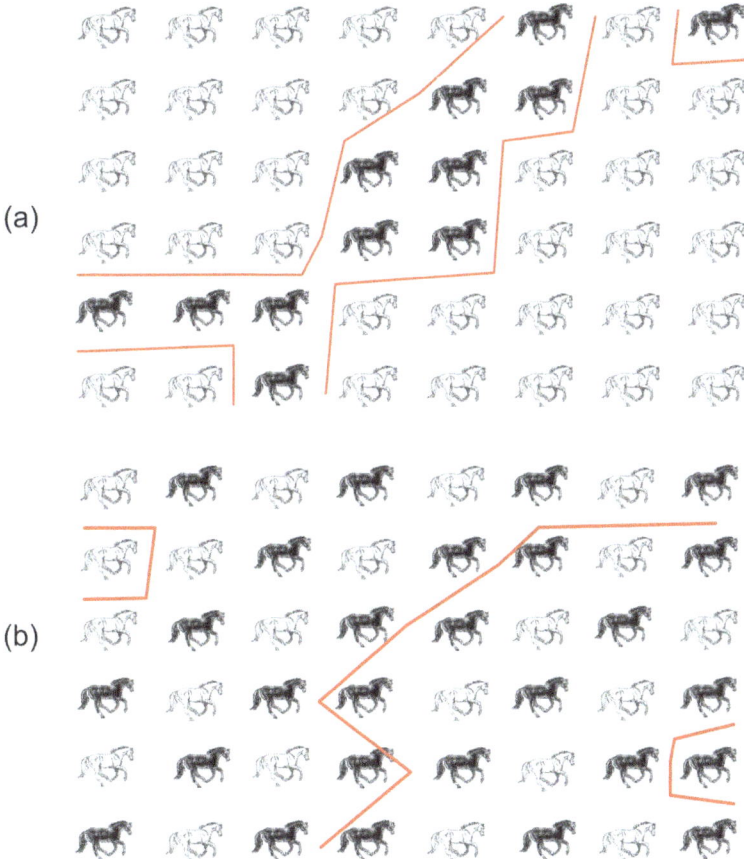

Figure 3.54. Short-range order in a crystal.

about the *average* crystal structure, whereas the diffuse intensity tells us about the *departures* from the average structure.

To understand this further, consider figure 3.54, where black and white horses represent two sorts of molecule or atom. In (a) you will note that there are regions with white horses and regions with black horses. So, clearly the long-range order has been broken, and the effect of this in diffraction space is to increase I_{diffuse} with respect to I_{Bragg}. This results in a diffuse broad peak sitting underneath the Bragg reflections. In (b) the correlated regions tend to have alternating black and white horses, so that there is a tendency to double the unit cell repeat distances. This then gives rise to a broad diffuse peak half-way between the reciprocal lattice nodes, i.e., between the Bragg peaks. In both cases, the width of the diffuse peaks is inversely proportional to the average sizes of the correlated domains. In the case where the horses are randomly distributed on the lattice sites, there will be a diffuse intensity adding to the overall background in the diffraction pattern. For fuller details on diffuse scattering, see the book by Welberry [25].

References

[1] Armstrong H E 1927 *Nature* **120** 478–9
[2] Ren Z, Bourgeois D, Helliwell J R, Moffat K, Šrajer V and Stoddard B L 1999 Laue crystallography: coming of age *J. Synchrotron Radiat.* **6** 891–917
[3] Clifton I J, Elder M and Hajdu J 1991 Experimental strategies in Laue crystallography *J. Appl. Crystallogr.* **24** 267–77
[4] Buras B, Gerward L, Glazer A M, Hidaka M and Staun Olsen J 1979 Quantitative structural studies by means of the energy-dispersive method with x-rays from a storage ring *J. Appl. Crystallogr.* **12** 531–6
[5] Ewald P P (ed) 1962 *Fifty Years of x-ray Diffraction* (Boston, MA: Springer)
[6] Harburn G, Taylor C and Welberry T R 1975 *Atlas of Optical Transforms* (Ithaca, NY: Cornell University Press)
[7] Mitchell D P and Powers P N 1936 Bragg Reflection of slow neutrons *Phys. Rev.* **50** 486–7
[8] von Halban H and Preiswerk P 1936 Experimental evidence of neutron diffraction *C.R. Hebd. Séances Acad.* **203** 73
[9] Grundy P J and Jones G A 1976 *Electron Microscopy in the Study of Materials* (London: Edward Arnold)
[10] Zou X, Hovmöller S and Oleynikov P 2011 *Electron Crystallography Electron Microscopy and Electron Diffraction* (Oxford: Oxford University Press)
[11] Giacovazzo C, Monaco H L, Artioli G, Viterbo D, Milanesio M, Gilli G, Gilli P, Zanotti G, Ferraris G and Catti M 2011 *Fundamentals of Crystallography* (Oxford: Oxford University Press)
[12] Ladd M and Palmer R 2003 *Structure Determination by x-ray Crystallography.* (New York: Springer)
[13] Clegg W 2015 *X-ray Crystallography* (Oxford: Oxford University Press)
[14] Fourier transforms: structure factors, phases and electron density [Online]. Available: www-structmed.cimr.cam.ac.uk/Course/Fourier/Fourier.html
[15] Read R J 1997 Model phases: probabilities and bias *Methods Enzymol.* **227** 110–28

[16] Glazer A M 2016 FOURIER2D and FOURIER3D: Programs to demonstrate Fourier synthesis in crystallography *J. Appl. Crystallogr.* **49** 6
[17] Chapuis G e-Crystallography [Online]. Available: http://escher.epfl.ch/eCrystallography/
[18] Oszlányi G and Sütő A 2004 Ab initio structure solution by charge flipping *Acta Crystallogr. Sect. A Found. Crystallogr.* **60** 134–41
[19] Rietveld H M 1969 A profile refinement method for nuclear and magnetic structures *J. Appl. Crystallogr.* **2** 65–71
[20] Tegel M, Hummel F, Su Y, Chatterji T, Brunelli M and Johrendt D 2010 Non-stoichometry and the magnetic structure of Sr_2CrO_3FeAs *Eur. Lett.* **10** 37006
[21] Sosnowska I, Przeniosło R, Schäfer W, Kockelmann W, Hempelmann R and Wysocki K 2001 Possible deuterium positions in the high-temperature deuterated proton conductor $Ba_3Ca_{1+y}Nb_{2-y}O_{9-\delta}$ studied by neutron and x-ray powder diffraction *J. Alloys Compd.* **328** 226–230
[22] Zhang N, Yokota H, Glazer A M, Ren Z, Keen D A, Keeble D S, Thomas P A and Ye Z-G 2014 The missing boundary in the phase diagram of $PbZr_{1-x}Ti_xO_3$ *Nat. Commun.* **5** 5231
[23] Glazer A M 2016 *Crystallography: A Very Short Introduction* (Oxford: Oxford University Press)
[24] Online Dictionary of Crystallography [Online]. Available: http://reference.iucr.org/dictionary/Main_Page
[25] Welberry T R 2004 *Diffuse x-ray Scattering and Models of Disorder* (Oxford: Oxford University Press)

IOP Concise Physics

A Journey into Reciprocal Space
A crystallographer's perspective
A M Glazer

Chapter 4

Dynamical diffraction

The electron is not as simple as it looks.

William Lawrence Bragg

4.1 Multiple scattering

So far, we have confined ourselves to cases where the kinematic approximation applies (only a small fraction of the incident radiation is diffracted). However, consider what can happen (figure 4.1) if the beam is scattered by planes in a perfect crystal, i.e., one without any defects or dislocations [1,2]. As usual, some of the incident beam can be thought of as being 'reflected' by the first plane it meets in accordance with Bragg's Law, whilst the remainder continues to the next plane. This too reflects the beam in the same way, but notice how this time it is also reflected back by the first plane as well. This continues with the next plane and so on, to give a series of multiple scatterings. Eventually, however, at some distance ξ (the *extinction distance*) within the crystal the beams become sufficiently weak that they no longer contribute significantly to the diffracted intensity.

The upshot of this is that our assumption that most of the incident radiation passes through the crystal and is not diffracted (first Born approximation) is not true when the crystal is perfect! This surprising result means that to use diffraction intensities to solve the crystal structure of a substance, contrary to what you might expect, crystallographers do not want perfect crystals.

Instead, the requirement is for what are called 'ideally imperfect' crystals: that is just enough imperfections to hinder the multiple scattering but not so much as to destroy the integrity of the crystal. Sometimes, crystallographers will deliberately drop a crystal into liquid nitrogen to introduce some mosaicity by thermally shocking it for this purpose. Even so, it is often found that the intensities of some reflections are less than those calculated theoretically from the crystal structure. This is known by crystallographers as *extinction* and it is often corrected for in crystallographic computer programs usually by empirical correction factors.

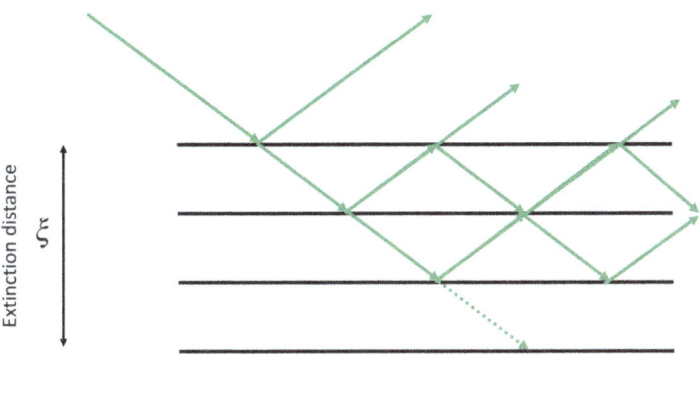

Figure 4.1. The effect of diffraction (reflection) by successive planes in a perfect crystal.

In transmission electron diffraction, multiple scattering is a more serious problem, since a very thin crystal is normally used in order to gain enough electron beam penetration, and this is usually therefore highly perfect. The result is that quite often the central beam maximum can appear weaker than some of the diffracted beams.

4.2 Renninger effect

Multiple scattering can be a serious problem in the determination of crystal structures by x-ray or neutron diffraction. A particular case arises when there are two reciprocal lattice points $h_1 k_1 l_1$ and $h_2 k_2 l_2$ lying on the Ewald sphere to give two diffracted beams at the same time. Thus, there are simultaneously two scattering vectors

$$\begin{aligned} \mathbf{k}_1 - \mathbf{k}_0 &= h_1 \mathbf{a}^* + k_1 \mathbf{b}^* + l_1 \mathbf{c}^* \\ \mathbf{k}_2 - \mathbf{k}_0 &= h_2 \mathbf{a}^* + k_2 \mathbf{b}^* + l_2 \mathbf{c}^* \end{aligned} \quad (4.1)$$

Subtracting one from the other

$$\mathbf{k}_1 - \mathbf{k}_2 = (h_1 - h_2)\mathbf{a}^* + (k_1 - k_2)\mathbf{b}^* + (l_1 - l_2)\mathbf{c}^* \quad (4.2)$$

This creates a third diffracted beam coming from $(h_1 - h_2, k_1 - k_2, l_1 - l_2)$ planes.

Figure 4.2 illustrates this. The incident beam is scattered simultaneously with wave-vectors $\mathbf{k}_1$ and $\mathbf{k}_2$. If the beam with wave-vector $\mathbf{k}_1$ now acts like an incident beam in the crystal a new Ewald sphere (dashed) can be drawn and then a third reciprocal lattice point is found on the surface at $h_1 - h_2, k_1 - k_2, l_1 - l_2$ given by

$$\mathbf{k}_3 = \mathbf{k}_1 - \mathbf{k}_2 \quad (4.3)$$

Generally, such extra so-called Renninger reflections tend to be weak but may appear in places where the space group symmetry dictates that there should be no intensity. It is therefore important in using diffraction data to recognize the existence of these extra reflections, when present, and remove them from the list of usable reflections (they can sometimes be recognized by being sharper than the normal

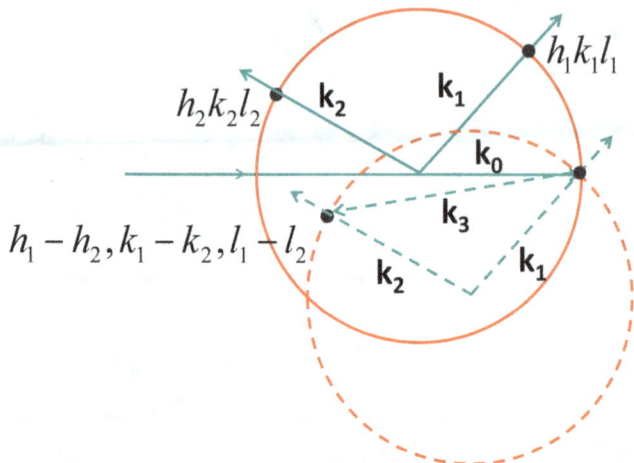

Figure 4.2. Renninger effect produced when two reciprocal lattice points lie simultaneously on the Ewald sphere.

Bragg reflections). In some cases, interference from Renninger effects can be a serious matter: for instance, in the structure determination of the material LiNbO$_3$ a total of 42 Renninger reflections was observed!

4.3 Two-beam approximation in electron diffraction

Problems with multiple scattering are of particular relevance to transmission electron diffraction, and the analysis of the diffraction conditions, where the crystal samples are very thin and likely to be nearly perfect. As a result, the understanding of contrast in electron microscope images is complicated (see [3, 4]). Note that the dynamical theory for electrons is similar to that for x-ray diffraction from perfect crystals.

One can gain some idea of how to approach the subject through considering a two-beam approximation to the dynamical theory. While this is not completely general, it still fits most situations quite well. Consider two electron beams interacting, i.e., two reciprocal lattice points lie on the Ewald sphere at the same time for a perfect crystal. We start by writing the electron plane wave function as

$$\psi(\mathbf{r}) = u_k e^{i\mathbf{k}\cdot\mathbf{r}} \qquad (4.4)$$

According to Bloch's theorem, we can then write

$$\psi(\mathbf{r}) = e^{i\mathbf{k}\cdot\mathbf{r}} \sum_{\mathbf{g}} C_g e^{i\mathbf{g}\cdot\mathbf{r}} \qquad (4.5)$$

The vector **g** is a reciprocal lattice vector and the summation is over all its possible values. For the two-beam case we take just two terms

$$\psi(\mathbf{r}) = C_0 e^{i\mathbf{k}\cdot\mathbf{r}} + C_g e^{i(\mathbf{k}+\mathbf{g})\cdot\mathbf{r}} \qquad (4.6)$$

Now the electron wave interacts with a periodic potential energy $V(\mathbf{r})$ in the crystal due to the presence of the atomic nuclei, and this can be written as a Fourier series:

$$V(\mathbf{r}) = \sum_g V_g e^{i\mathbf{g}\cdot\mathbf{r}} \qquad (4.7)$$

We now write

$$U_g = \frac{2m_e}{\hbar^2} V_g \qquad (4.8)$$

and

$$K^2 = \frac{2m_e}{\hbar^2}(E - V_0) \qquad (4.9)$$

E is the energy of the incident electrons and K is the electron wave-vector taking into account the refractive index of the crystal and is related to the wave-vector K_{vac} *in vacuo* by

$$K^2 + \frac{2m_e}{\hbar^2} V_0 = \frac{2m_e}{\hbar^2} E = K^2_{vac} \qquad (4.10)$$

Then, applying the time-independent Schrödinger equation

$$\frac{\hbar^2}{2m_e}\nabla^2\psi + V(\mathbf{r})\psi = E\psi \qquad (4.11)$$

we get

$$\frac{\hbar^2}{2m_e}\left[-\mathbf{k}^2 C_0 e^{i\mathbf{k}\cdot\mathbf{r}} - (\mathbf{k}+\mathbf{g})^2 C_g e^{i(\mathbf{k}+\mathbf{g})\cdot\mathbf{r}}\right] + \sum_g V_g e^{i\mathbf{g}\cdot\mathbf{r}}\left[C_0 e^{i\mathbf{k}\cdot\mathbf{r}} + C_g e^{i(\mathbf{k}+\mathbf{g})\cdot\mathbf{r}}\right]$$
$$= E\left[C_0 e^{i\mathbf{k}\cdot\mathbf{r}} + C_g e^{i(\mathbf{k}+\mathbf{g})\cdot\mathbf{r}}\right] \qquad (4.12)$$

Comparing coefficients of $e^{i\mathbf{k}\cdot\mathbf{r}}$ and $e^{i(\mathbf{k}+\mathbf{g})\cdot\mathbf{r}}$ on both sides of this equation, we then obtain two simultaneous equations:

$$\frac{\hbar^2}{2m_e}\mathbf{k}^2 C_0 + V_0 C_0 + V_{-g} C_g = E C_0$$
$$\frac{\hbar^2}{2m_e}(\mathbf{k}+\mathbf{g})^2 C_g + V_0 C_g + V_g C_0 = E C_g \qquad (4.13)$$

Rearranging and using equations (4.8) and (4.9)

$$(K^2 - \mathbf{k}^2)C_0 + U_{-g} C_g = 0$$
$$U_g C_0 + (K^2 - |\mathbf{k}+\mathbf{g}|^2)C_g = 0 \qquad (4.14)$$

For a non-trivial solution

$$\begin{vmatrix} K^2 - \mathbf{k}^2 & U_{-g} \\ U_g & K^2 - |\mathbf{k} + \mathbf{g}|^2 \end{vmatrix} = 0 \qquad (4.15)$$

and so

$$(K^2 - \mathbf{k}^2)(K^2 - |\mathbf{k} + \mathbf{g}|^2) = U_g U_{-g} \qquad (4.16)$$

As K, $\mathbf{k}$ and $|\mathbf{k} + \mathbf{g}|$ are very large compared with the differences between them (this is because the periodic potential is much smaller than that of the electron beam), this equation becomes

$$(\mathbf{k} - K)(|\mathbf{k} + \mathbf{g}| - K) = \frac{U_g U_{-g}}{4K^2} = \frac{|U_g|^2}{4K^2} \qquad (4.17)$$

It can be seen that the result is a quadratic equation in K, showing that there are two solutions. Consider the special case $\mathbf{k} = \mathbf{k} + \mathbf{g}$ where a standing solution occurs. Then equation (4.16) becomes

$$(\mathbf{k}^2 - K^2) = |U_g|^2 \qquad (4.18)$$

Solving for the roots K_+ and K_-

$$K_+^2 - K_-^2 = 2U_g \qquad (4.19)$$

$$(K_+ + K_-)(K_+ - K_-) = 2U_g \qquad (4.20)$$

and then

$$\Delta K = (K_+ - K_-) = \frac{2U_g}{2K} = \frac{2m_e V_g}{\hbar^2 K} = \xi^{-1} \qquad (4.21)$$

Definition. *The quantity ξ is known as the extinction distance and can be thought of as a mean free path for the electron wave.*

Therefore, the crystal splits the incident beam into two Bloch waves, each with different wave-vectors. This means that there will be phase differences between the two waves and hence they can interfere with each other to produce interference fringes. The crystal therefore acts as its own interferometer.

In order to understand this better, consider figure 4.3 which shows the situation for the kinematic case. The full circle is the usual Ewald sphere for which the wave-vector $\mathbf{k}_0$ represents the incident wave and $\mathbf{k}$ the diffracted wave. Now in the dynamical theory the two beams share energy rather as happens in a balanced pendulum, since in a perfect crystal both may be intense waves within the crystal. If we construct the loci of the centres of the Ewald sphere around the two reciprocal lattice nodes O and P the dashed spheres are obtained.

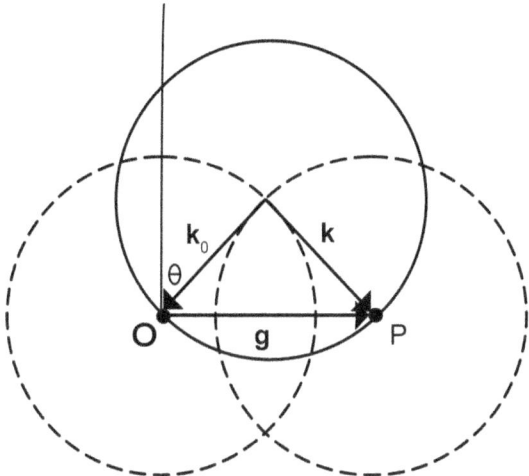

Figure 4.3. Ewald sphere in the kinematic approximation.

Now in the two-beam dynamical case, for high electron energies, equation (4.17) defines an energy *dispersion surface* that has two branches asymptotic to the spherical surfaces (figure 4.4) separated at the centre by $\Delta K = \xi^{-1}$.

This means that instead of a single wave-vector to the point O, as in the kinematic approximation, there are two wave-vectors $\mathbf{K}_1$ and $\mathbf{K}_2$ one from each branch, and similarly $\mathbf{K}_1 + \mathbf{g}$ and $\mathbf{K}_2 + \mathbf{g}$ to the point P. The central line BZ, denotes the so-called Brillouin zone boundary corresponding to the Bragg reflection condition. Wave-vectors coming from the two points on the upper and lower branches on BZ therefore signify standing-wave solutions. Note that as the extinction distance increases the two branches become close together until they finally touch and the kinematic situation is recovered (in this case the electron can be represented by a single plane wave of wave-vector $\mathbf{k}$ or $\mathbf{k} + \mathbf{g}$. The extinction distance is therefore integrally linked to breakdown of the first Born approximation.

Figure 4.5 shows the energy surfaces around the reciprocal lattice points in a repeated zone scheme where vectors $\mathbf{k}$ and $\mathbf{k} + \mathbf{g}$ are equivalent by translational symmetry. The total wave-functions for the two solutions are given by

$$\psi_1 = C_0(1)e^{i\mathbf{K}_1 \cdot \mathbf{r}} + C_g(1)e^{i(\mathbf{K}_1 + \mathbf{g}) \cdot \mathbf{r}}$$
$$\psi_2 = C_0(2)e^{i\mathbf{K}_2 \cdot \mathbf{r}} + C_g(2)e^{i(\mathbf{K}_2 + \mathbf{g}) \cdot \mathbf{r}} \qquad (4.22)$$

In the Bragg reflection position we can express the wave-functions for the two beams as

$$\psi_1 = \frac{1}{\sqrt{2}} C \{ e^{i\mathbf{K}_1 \cdot \mathbf{r}} - e^{i(\mathbf{K}_1 + \mathbf{g}) \cdot \mathbf{r}} \}$$
$$\psi_2 = \frac{1}{\sqrt{2}} C \{ e^{i\mathbf{K}_2 \cdot \mathbf{r}} - e^{i(\mathbf{K}_2 + \mathbf{g}) \cdot \mathbf{r}} \} \qquad (4.23)$$

If we take $\mathbf{K}_1 = \mathbf{K}_2 = -\mathbf{g}/2$

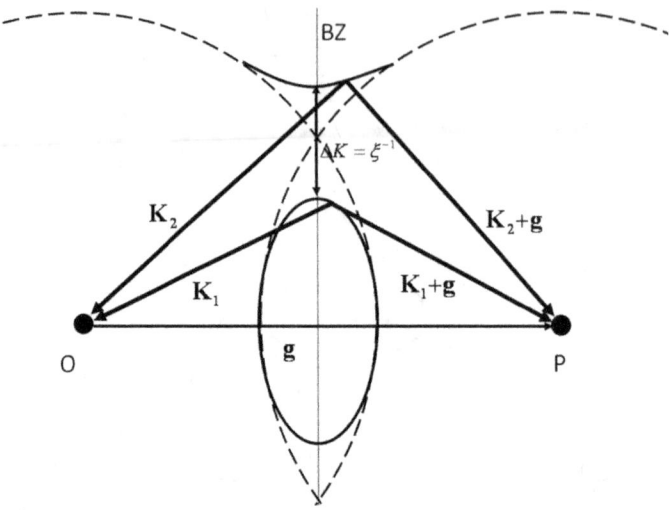

Figure 4.4. Dispersion surface for two-beam dynamical diffraction.

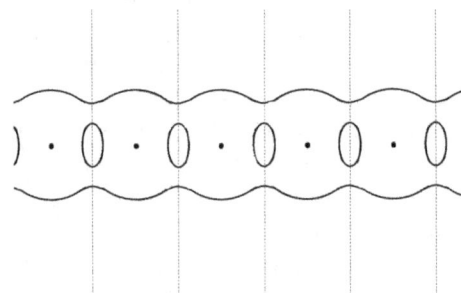

Figure 4.5. Energy dispersion surface in repeated zone scheme.

$$\psi_1 = \sqrt{2}\,iC \sin \frac{1}{2}\mathbf{g}\cdot\mathbf{r}$$
$$\psi_2 = \sqrt{2}\,C \cos \frac{1}{2}\mathbf{g}\cdot\mathbf{r} \qquad (4.24)$$

and then

$$\psi_1\psi_1^* = |\psi_1|^2 = 2C^2 \sin^2\left(\frac{1}{2}\mathbf{g}\cdot\mathbf{r}\right)$$
$$\psi_2\psi_2^* = |\psi_2|^2 = 2C^2 \cos^2\left(\frac{1}{2}\mathbf{g}\cdot\mathbf{r}\right) \qquad (4.25)$$

We can get an idea of what this means by considering each of the waves with respect to the arrangement of atoms (figure 4.6).

For wave (1) the maxima are *between* the atomic planes, whereas for wave (2) they lie *on* the atomic planes. Both Bloch waves have the same total energy, but since electrons in wave (2) are concentrated in a region of low potential energy they must

A Journey into Reciprocal Space

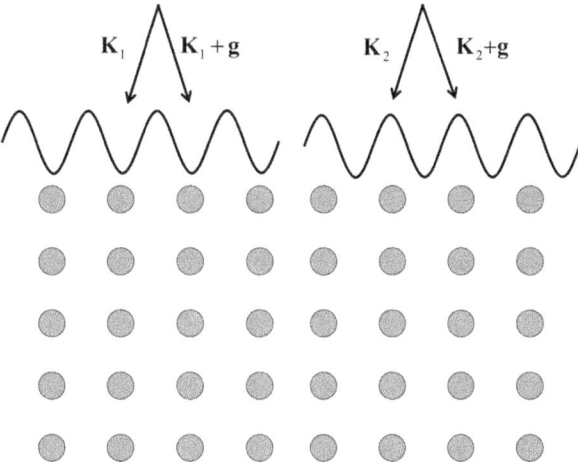

Figure 4.6. Schematic diagram of the two types of wave field at the Bragg reflecting position. The current flow vector is parallel to the reflecting planes.

on average have a higher kinetic energy than electrons in wave (1). It is this difference that is the reason for the separation of the dispersion branches by ΔK. It also follows that wave (2) is scattered more than wave (1) because it is concentrated close to the atoms where there is a greater likelihood of interaction with the atoms.

We note in passing that this treatment is closely similar to that used in electronic band theory, the main difference being that here the energy is fixed and one solves for the wave-vectors, whereas there it is the other way around, with the wave-vectors fixed and the Schrödinger equation solved for energy solutions.

4.4 Pendellösung or thickness fringes

As an example of how the two-beam theory of electron diffraction can be used, consider a wedge-shaped crystal inside an electron microscope (figure 4.7).

For thickness t, the amplitude of the directly transmitted beam through the crystal is given by

$$\psi_0(t) = C_0(1)e^{iK_1 t} + C_0(2)e^{iK_2 t} \qquad (4.26)$$

and for the diffracted beam

$$\psi_g(t) = C_g(1)e^{i(K_1+g)t} + C_g(2)e^{i(K_2+g)t} \qquad (4.27)$$

At the Bragg reflecting position

$$\mathbf{K} = \mathbf{K} + \mathbf{g} \qquad (4.28)$$

and

$$\begin{aligned} C_0(1) &= C_0(2) \\ C_g(1) &= C_g(2) \end{aligned} \qquad (4.29)$$

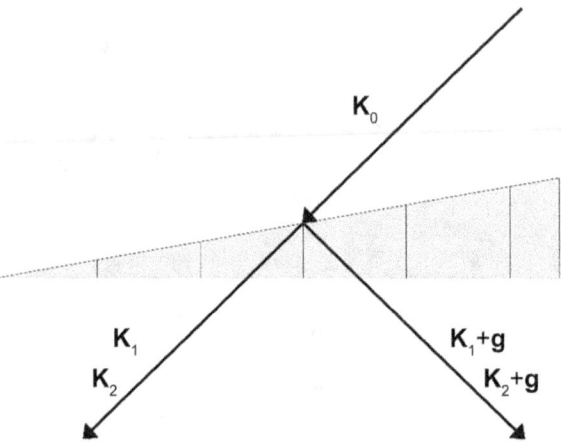

Figure 4.7. Electron beam transmitted and diffracted by a wedge-shaped crystal.

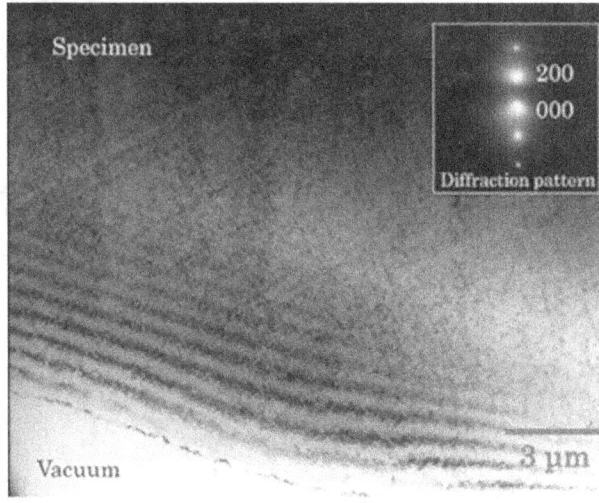

Figure 4.8. Thickness fringes shown by a wedge-shaped crystal of AlCu alloy in an electron microscope image [5]. Reproduced with permission of JEOL www.jeol.co.jp.

and so the intensity for the transmitted beam will be given by
$$I_{trans} = \psi_0(t)\psi_0^*(t) \propto 1 + 1 + \cos(K_1 - K_2)t \qquad (4.30)$$
Therefore
$$I_0 \propto \cos^2\left[\frac{K_1 - K_2}{t}\right] \qquad (4.31)$$
Similarly, for the diffracted beam
$$I_g \propto \sin^2\left[\frac{K_1 - K_2}{t}\right] \qquad (4.32)$$

and then

$$I_0 + I_g = \text{const} \qquad (4.33)$$

Therefore, the effect of this is to create in the electron microscope image of the wedge-shaped crystal a set of contrast fringes both in the image due to the direct beam and the image due to the diffracted beam. The periodicity of the fringes can then, in principle, be used to estimate the angle of the wedge-shaped crystal. Note that in practice, inelastic scattering processes cause absorption so that the intensity tends to decay with thickness t. An example of thickness fringes is shown in figure 4.8.

References

[1] Authier A 2003 *Dynamical Theory of x-ray Diffraction* (Oxford: Oxford University Press)
[2] Als-Nielsen J and McMorrow D 2011 *Elements of Modern x-ray Physics* (New York: Wiley)
[3] Grundy P J and Jones G A 1976 *Electron Microscopy in the Study of Materials* (London: Edward Arnold)
[4] Hirsch P, Howie A, Nicholson R, Pashley D W and Whelan M J 1977 *Electron Microscopy of Thin Crystals* (Malabar, FL: Krieger Pub)
[5] www.jeol.co.jp/en/words/emterms/search_result.html?keyword=equal%20thickness%20fringe

IOP Concise Physics

A Journey into Reciprocal Space
A crystallographer's perspective
A M Glazer

Chapter 5

Waves in a periodic medium

Physicists use the wave theory on Mondays, Wednesdays and Fridays, and the particle theory on Tuesdays, Thursdays and Saturdays.

William Henry Bragg

5.1 Waves in space

We shall now look at the way in which reciprocal space can be used in order to describe the quantum states for various waves travelling in a periodic medium, i.e., in a crystal. These waves arise through elementary excitations such as those for electrons or phonons. We have already seen how the incidence of x-ray, neutron and electron beams leads to diffraction by the crystal and how that can be described through the Ewald construction applied to the reciprocal lattice. In much the same way, wave states for elementary excitations can also be included, and indeed such waves can be considered to be scattered or diffracted by the underlying lattice potential. For this reason, in the case of elastic scattering, where there is no difference in energy between incident and diffracted waves, the corresponding wave-vectors are related by:

$$k_0 = k \tag{5.1}$$

In addition, the Ewald construction has shown that

$$\mathbf{k} - \mathbf{k}_0 = \mathbf{g} \tag{5.2}$$

is the condition for a wave to be diffracted, or turning this around the scattered wave vector **k** is

$$\mathbf{k} = \mathbf{k}_0 + \mathbf{g} \tag{5.3}$$

In terms of conservation of wave momentum, the scattered momentum is

$$\hbar \mathbf{k} = \hbar \mathbf{k}_0 + \hbar \mathbf{g} \tag{5.4}$$

Notation. *Reciprocal space is sometimes called k-space or momentum space.*

Correspondingly, for elastic scattering the energy between the incident and scattered waves is also conserved:

$$\hbar\omega_0 = \hbar\omega \tag{5.5}$$

5.2 Periodic boundary conditions

We now consider how to specify the quantum states for waves travelling in a crystal, bearing in mind that waves or elementary excitations corresponding to electrons and phonons have a multitude of wavelengths with scales varying from the crystal dimensions down to nanometers. So, the problem that needs to be addressed is: how can one fit all these waves within the confines of the crystal? Now one way to do this is to recognize that for most purposes the surfaces of the crystal are so far away with respect to unit cell dimensions that we can treat the crystal as if it were infinite in extent. Given this, we can then, without loss of generality, treat the problem mathematically as if one end of the crystal, as described by a lattice, at ∞ is connected to the other at −∞. This tying together of opposite ends of the lattice is known as a *Periodic Boundary Condition*. The simplest way to see this is to take a one-dimensional example.

Figure 5.1 illustrates this with a simple example of 12 lattice points spaced *a* apart and arranged on a circle of circumference L (to describe a real crystal of normal size more like 10^{25} lattice points would be needed). The idea now is to consider possible wave motions that will displace these lattice points from their ideal positions (a mechanical analogue would be to have a circular string of atoms set into vibrational motion). Furthermore, any wave around this circular string must be continuous and not have any discontinuities. The fundamental wave will then have a wavelength

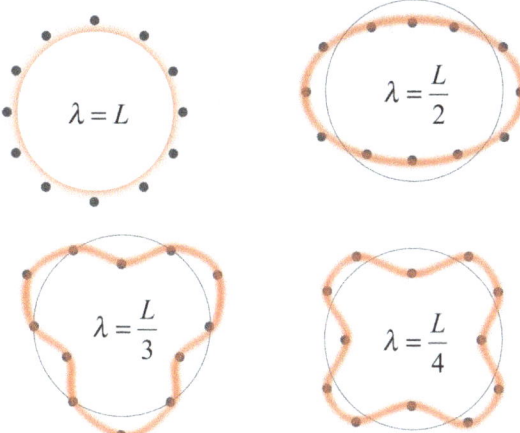

Figure 5.1. Examples of different waves for a circular array of lattice points of spacing *a*.

equal to the circumference of the circle and the first harmonic will have a wavelength equal to half the circumference, and so on. The wavelengths therefore are

$$L \quad \frac{L}{2} \quad \frac{L}{3} \quad \frac{L}{4} \quad \frac{L}{5} \quad \frac{L}{6} \quad \cdots\cdots \quad (5.6)$$

Now if, instead, we write this sequence in terms of wave-vectors, we get the following nice sequence:

$$\frac{2\pi}{L} \quad \frac{4\pi}{L} \quad \frac{6\pi}{L} \quad \frac{8\pi}{L} \quad \frac{10\pi}{L} \quad \frac{12\pi}{L} \quad \cdots\cdots \quad (5.7)$$

The advantage of specifying the waves by wave-vectors rather than by wave-lengths should be apparent: we end up with a series of points in k space each equally separated by a fixed reciprocal distance of $2\pi/L$. Furthermore, because L is so much bigger than the unit cell distance, we can see that on the scale of the reciprocal lattice these points are extremely close together compared with the separation of the reciprocal lattice points, and for most purposes can be considered to form a quasi-continuum.

Figure 5.2 illustrates this schematically, where

$$\frac{2\pi}{L} \ll \frac{2\pi}{a} \quad (5.8)$$

It is obvious that this can be repeated in three dimensions for a crystal of dimensions $L \times L \times L$. The result is that the whole of reciprocal space is filled with a fine mesh of points representing all the quantum states for elementary excitations in the crystal. Figure 5.3 shows what this looks like in two dimensions; an orthogonal reciprocal lattice has been drawn here for simplicity.

You will notice that the wave states have been marked in all over reciprocal space. This is because the solutions to the wave equation will be periodic in the reciprocal lattice. We can see this as follows. Consider a plane wave and subtract a reciprocal lattice vector $\mathbf{g}$ from the wave-vector:

$$e^{i(\mathbf{k}-\mathbf{g})\cdot\mathbf{t}_n} = e^{i\mathbf{k}\cdot\mathbf{t}_n} e^{-i\mathbf{g}\cdot\mathbf{t}_n} \quad (5.9)$$

In the figure the reciprocal lattice vector has been drawn with its origin displaced in order to complete a triangle of vectors. Now, recall that

$$\mathbf{g} = h\mathbf{a}^* + k\mathbf{b} ++ l\mathbf{c}^* \quad (5.10)$$

and

$$\mathbf{t}_n = n_1\mathbf{a} + n_2\mathbf{b} + n_3\mathbf{c} \quad (5.11)$$

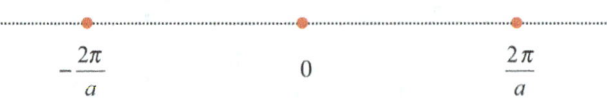

Figure 5.2. One-dimensional reciprocal lattice (red points) space $2\pi/a$ and wave-vectors (black points) space $2\pi/L$ apart.

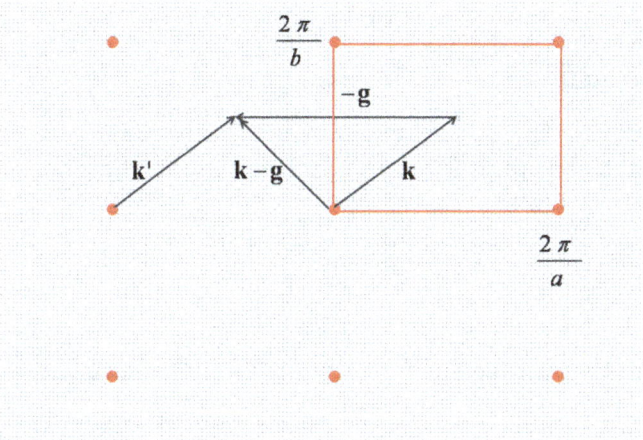

Figure 5.3. Two-dimensional reciprocal lattice (red points) spaced $2\pi/a$ by $2\pi/b$ and wave-vectors (black points) spaced $2\pi/L$ by $2\pi/L$. The red box is an example of a primitive unit cell in the reciprocal lattice.

In other words, the wave solutions are invariant under addition or subtraction of integer multiples of the reciprocal vector **g**. In figure 5.3 we see that subtracting the vector **g** from the wave-vector **k** from the origin (0, 0) leads to the vector **k** − **g** pointing to the left with respect to our chosen origin in the reciprocal lattice, but it is also equivalent to the wave-vector **k**′ pointing to the right with respect to an origin chosen at

$$\left(-\frac{2\pi}{a}, 0\right)$$

This illustrates an important idea. It shows that because the reciprocal lattice is translationally invariant, all wave solutions also will be translationally invariant. This is equivalent to saying that in real space the crystal structure is translationally invariant. It therefore means that we can define all the wave states within a unit cell in reciprocal space and then all other reciprocal unit cells will contain the same wave states. The reciprocal lattice behaves just like a real lattice in that it acts as a template for translational symmetry. In real space we have atoms inside unit cells, but in reciprocal space we have wave states inside reciprocal space unit cells.

5.3 Bloch's theorem

Consider the effect of an operator $\hat{T}$ acting on the function $e^{i\mathbf{k}\cdot\mathbf{r}}$.

$$\hat{T}e^{i\mathbf{k}\cdot\mathbf{r}} = e^{i\hat{T}\mathbf{k}\cdot\mathbf{r}} \qquad (5.12)$$

In addition, consider a scalar product. Because this gives a scalar as a result it is not affected by the operator $\hat{T}$, e.g.

$$\mathbf{p} \cdot \mathbf{q} = \hat{T}\mathbf{p} \cdot \hat{T}\mathbf{q} \qquad (5.13)$$

Therefore

$$\mathbf{k} \cdot \hat{T}^{-1}\mathbf{r} = \hat{T}(\mathbf{k}) \cdot \hat{T}(\hat{T}^{-1}\mathbf{r}) = \hat{T}\mathbf{k} \cdot \mathbf{r} \tag{5.14}$$

In general, we can therefore write

$$\hat{T}f(\mathbf{r}) = f(\hat{T}^{-1}\mathbf{r}) \tag{5.15}$$

Suppose now that the operator $\hat{T}$ is in fact the translation operator t_n. Then we can write

$$\begin{aligned}\hat{T}e^{i\mathbf{k}\cdot\mathbf{r}} &= e^{i\mathbf{k}\cdot(\mathbf{r}\pm\mathbf{t}_n)} \\ &= e^{i\mathbf{k}\cdot\mathbf{r}}e^{\pm i\mathbf{k}\cdot\mathbf{t}_n} \\ &= e^{i\mathbf{k}\cdot\mathbf{r}}\end{aligned} \tag{5.16}$$

Therefore, suitable basis functions for translation can be taken as $e^{i\mathbf{k}\cdot\mathbf{r}}$.

Definition. *Bloch's theorem states that the wave function for a particle moving in a periodic potential $V(\mathbf{r})$ with periodicity t_n is given by*

$$\psi_k(\mathbf{r}) = u_k(\mathbf{r})e^{i\mathbf{k}\cdot\mathbf{r}} \tag{5.17}$$

5.4 Brillouin zones

As explained above, for waves propagating in a periodic medium there are always other solutions to the wave-equation. Suppose we have waves described by the following wave-vectors

$$\begin{aligned}\mathbf{k}_1 &= \frac{\mathbf{g}}{2} \\ \mathbf{k}_2 &= -\frac{\mathbf{g}}{2} \\ \therefore k_1 &= k_2 = \frac{g}{2}\end{aligned} \tag{5.18}$$

Another way of expressing this is by

$$k_1 = \frac{2\pi}{\lambda} = \frac{g_{hkl}}{2} \tag{5.19}$$
$$\therefore 2d_{hkl} = \lambda$$

which is simply Bragg's law where $2\theta = 180°$. This means that the wave k_1 is reflected back to k_2 and vice versa. In other words, with respect to the chosen origin in reciprocal space, there are two waves of equal wave-vector magnitude travelling in opposite directions, i.e., a standing wave consisting of two opposite real components. Such a pair of waves is said to end on the so-called Brillouin zone boundary. We met this earlier when discussing the two-beam approximation in dynamical scattering.

What about waves with wave-vectors that do not end on the Brillouin zone boundary? There are still other solutions to the wave-equation. Suppose one is for $\mathbf{k}$ and the other for $\mathbf{k} - \mathbf{g}$. Referring back to figure 5.3, with the origin at the centre of the diagram the vector $\mathbf{k}$ represents a real wave solution within the specified

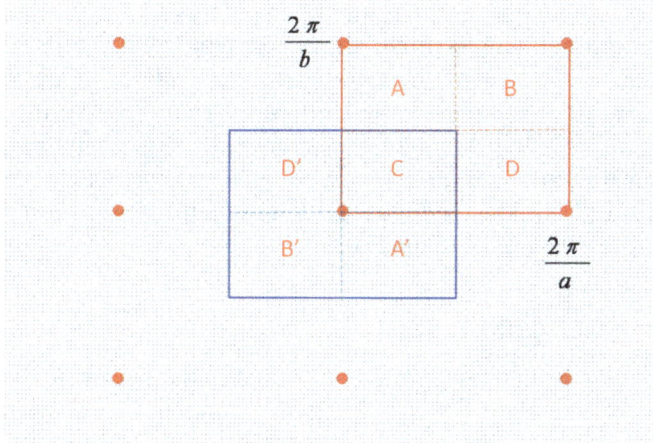

Figure 5.4. Two-dimensional reciprocal lattice with two unit cells.

reciprocal unit cell with a *component* travelling to the right, whereas **k** − **g** is for a wave whose wave-vector lies outside the reciprocal unit cell: this corresponds to a solution with a *component* travelling to the left. The two wave solutions in figure 5.3 differ in wave-vector by an amount **g**. Thus, in terms of wave-vector components $\mathbf{k}_1$ and $\mathbf{k}_2$ parallel to **g** we can write

$$\mathbf{k}_1 - \mathbf{g} = \mathbf{k}_2 \qquad (5.20)$$

Therefore

$$\frac{2\pi}{\lambda_1} - \frac{2\pi}{a} = \frac{2\pi}{\lambda_2}$$
$$\therefore a = \frac{\lambda_1 \lambda_2}{\lambda_2 - \lambda_1} \qquad (5.21)$$

where a is the unit cell repeat distance.

Definition. *The Brillouin zone is a unit cell in reciprocal space.*

Now this definition may surprise you for it is not the way it is usually defined in most textbooks, but that is because it seems to be poorly understood by many physicists and chemists. The point about this definition is that the Brillouin zone is a region of reciprocal space that contains *all* the allowed wave states. This can be compared with the way a real unit cell contains all the atoms needed to describe the whole crystal. What is often not realized is that the Brillouin zone can be constructed in an infinite number of ways in reciprocal space, because like all unit cells it can be drawn in any way we like: it just has to satisfy the notion that when repeated it fills all (reciprocal) space[1][2].

Let's look at a particular construction (figure 5.4). The red box is what we may call a conventional reciprocal unit cell with lattice points at each corner.

Because there is only one lattice point in this cell it is primitive. Now the blue cell is the same unit cell but with its origin displaced so that a lattice point now lies at its centre. The regions ABC and D together contain all the allowed wave states (property of a reciprocal unit cell) and their counterparts can be found in the blue cell as A' B' C and D'. Notice that this too, according to our definition, is a Brillouin zone. The only difference is in the way we have defined the origin and axes. However, instead of thinking of the blue cell as a simple translation of the red cell, there is yet another way of obtaining it.

5.5 Wigner–Seitz cell

The Wigner–Seitz cell (sometimes referred to as a *proximity* cell, or *Dirichlet* domain or a *Voronoi* cell) is a unit cell in a lattice that is constructed in a particular way. It can be used in real space, although with not much practical use for our purposes here, as well as in reciprocal space. To carry out the construction

1. Choose a lattice point
2. Draw vectors from this lattice point to all other lattice points
3. Bisect each of these vectors by planes *perpendicular* to the vectors
4. The Wigner–Seitz cell is the smallest enclosed volume around the initial lattice point.

It can be seen that the blue unit cell in figure 5.4 can be obtained by this procedure.

Suppose now we consider a centred lattice. For example, take a cubic body-centred lattice. Figure 5.5 shows the various stages in constructing the Wigner–Seitz cell. The result is something that looks very different from the parallelepiped-shaped unit cells that we have used so far. Nevertheless, it is a unit cell, as when stacked together with identical copies it fills all space without leaving gaps, the requirement for being a general unit cell. This can be seen in figure 5.6.

So, the question that has to be asked is: why use such a peculiar construction in order to simply describe wave states in reciprocal space? The answer to this is that, first of all, because it is constructed around a single lattice point, the Wigner–Seitz cell is always a primitive unit cell and therefore the smallest volume possible. This means that if we are counting quantum states, we need only do this once, whereas if we use a conventional centred unit cell we would then be counting the same states several times over. The second advantage of this construction is that the unit cell produced when looked at in isolation readily shows the full symmetry of the

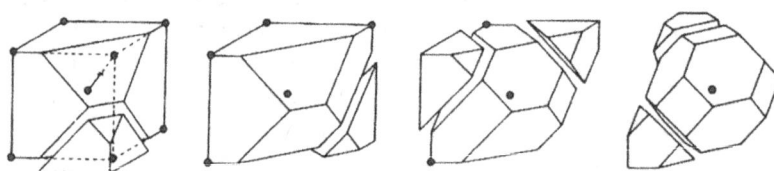

Figure 5.5. The construction of a Wigner–Seitz cell for a cubic I-lattice. The lattice points at the origin and neighbouring sites are shown.

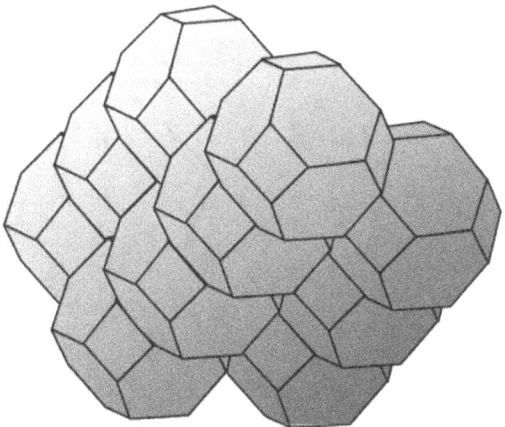

Figure 5.6. The stacking of Wigner–Seitz cells for a body-centred cubic lattice.

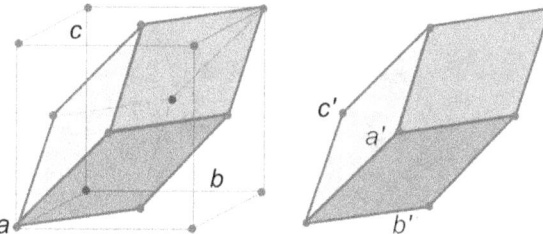

Figure 5.7. An example of the construction of a parallelepiped-shaped primitive unit cell out of an all-face-centred cubic unit cell.

underlying lattice from which it has been constructed. In the example above, we can immediately see in the shape of the cell that there are four 3-fold axes of symmetry, thus showing that it has been derived within a cubic lattice. Suppose instead of using a Wigner–Seitz construction we defined a primitive parallelepiped-shaped cell. For example, consider in figure 5.7 a cubic all-face-centred unit cell. The shaded cell is one of the infinite number of primitive unit cells that can be constructed with this lattice and, because it is primitive, it has one quarter of the volume of the F-cell.

The relationship between the two choices of unit cell is given by the equation

$$\begin{bmatrix} a' \\ b' \\ c' \end{bmatrix} = \begin{bmatrix} 1/2 & 0 & 1/2 \\ 1/2 & 1/2 & 0 \\ 0 & 1/2 & 1/2 \end{bmatrix} \begin{bmatrix} a \\ b \\ c \end{bmatrix} \qquad (5.22)$$

The primitive cell is shown on the right-hand side in isolation. The problem is that if you now hand this unit cell to someone and ask to which crystal system does this belong, the likely answer will be trigonal. This is because it is easy to spot a 3-fold axis of symmetry along the long diagonal and virtually impossible to locate the remaining three 3-fold axes necessary to define a cubic crystal system. In fact, the angle α between the axes can be found from (5.22) using

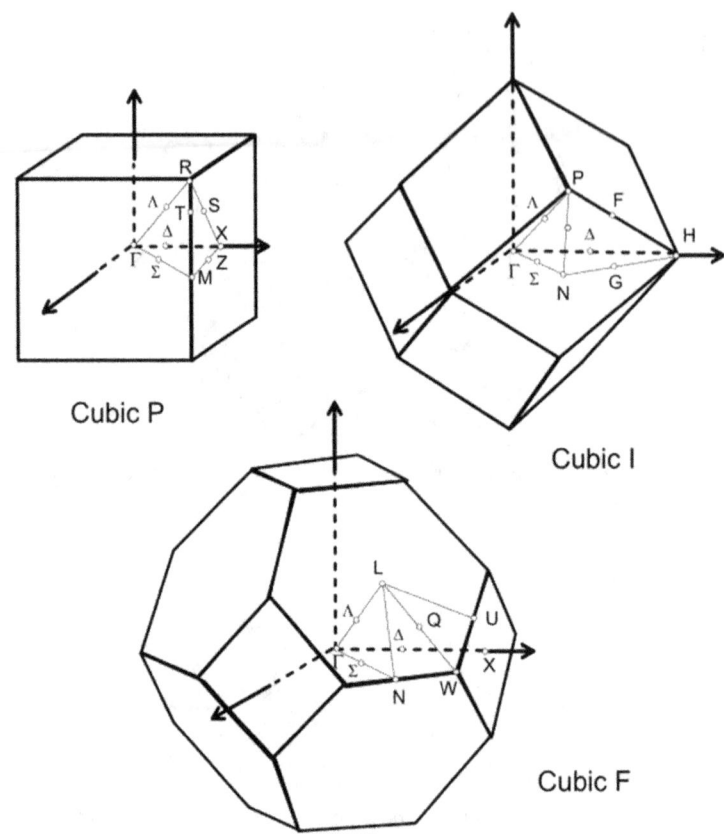

Figure 5.8. Brillouin zones using Wigner–Seitz constructions for cubic lattices.

$$\frac{1}{2}(\mathbf{a}+\mathbf{c})\cdot\frac{1}{2}(\mathbf{a}+\mathbf{b}) = \frac{1}{4}a^2 = \left|\frac{1}{2}(\mathbf{a}+\mathbf{c})\right|\left|\frac{1}{2}(\mathbf{a}+\mathbf{b})\right|\cos\alpha \qquad (5.23)$$

to be exactly 60°. Therefore, this rhombohedral shape is a very special one, but simply by looking at it we cannot be sure if α is exactly 60° without very careful measurement. The Wigner–Seitz cell for the F-centred cubic lattice is shown in figure 5.8, where the four 3-fold axes can be seen.

Thus, the Wigner–Seitz unit cell in reciprocal space is a convenient container for all the allowed wave states and so is a useful way of describing a Brillouin zone. However, and this is the point that is not properly understood by many working scientists, while this is fine for high-symmetry systems, the Wigner–Seitz construction becomes complicated for low symmetries, such as in monoclinic and triclinic lattices. This was first pointed out by Bradley and Cracknell [1]. The exact shapes of the Wigner–Seitz cells depend then on the ratios of the unit cell axes and angles, and it is easy to make mistakes in trying to draw them (see appendix). These days computer programs can handle the Wigner–Seitz shapes even for low-symmetry lattices, and so this concern is less serious than in earlier times. Unfortunately, the majority of text-books *define* the Brillouin zone *as* the Wigner–Seitz cell.

Table 5.1. Symbols for some critical points in the primitive cubic Wigner–Seitz unit cell.

Symbol	Fraction	Coordinate
Γ	0 0 0	0 0 0
R	½ ½ ½	$\pi/a\ \pi/a\ \pi/a$
M	½ ½ 0	$\pi/a\ \pi/a\ 0$
X	½ ½ 0	$\pi/a\ \pi/a\ 0$

Note that labels have been given to various special points in the Wigner–Seitz drawings. These are known as *critical points* because they occur at certain special positions in Brillouin zones defined by the Wigner–Seitz cells. There is unfortunately no internationally accepted consensus on their use, although the ones shown here are the most commonly used these days. Table 5.1 is a list of some of the more commonly used critical points in the primitive cubic system (the symbol Γ is universally accepted to represent the centre of the Brillouin zone for all crystal systems):

Conclusion. *The main reason for using a Wigner–Seitz unit cell in reciprocal space is that it is always a primitive unit cell and it readily shows the full symmetry of the underlying lattice.*

5.6 Higher-order Brillouin zones

Brillouin zones defined by the Wigner–Seitz construction as described above are often referred to as *first Brillouin zones*, for reasons that will become apparent when we later discuss how to treat elementary excitations. However, sometimes one meets so-called second, third and so on Brillouin zones. Their construction simply follows the same procedure as for Wigner–Seitz first Brillouin zones, except that the construction is extended further out in reciprocal space. We proceed by example to demonstrate how this is done for the simple case of a square reciprocal lattice.

Figure 5.9 shows how to construct the first Brillouin zone by drawing in vectors to the reciprocal lattice points at $\pm 2\pi/a$ and. $\pm 2\pi/b$. These lines are then bisected by perpendicular lines (planes in a three-dimensional lattice) running through $\pm\pi/a$ and. $\pm\pi/b$ to contain a square region (shown in blue). As explained above, this is a primitive unit cell in reciprocal space and contains all the allowed wave-vectors for the elementary excitations.

In figure 5.10, vectors are now drawn to the reciprocal lattice points at ($\pm 2\pi/a$, $\pm 2\pi/b$). These too are bisected perpendicularly to enclose not only the first Brillouin zone but also a region marked in pink. Note that the pink region has the same area (volume in three dimensions) as the first Brillouin zone and so by translational symmetry must also contain the same number of wave-states.

Figure 5.11 continues this process by drawing in the vectors to $2\pi/a$ and $\pm 2\pi/b$ and to $4\pi/a$ and $\pm 4\pi/b$. The lines that perpendicularly bisect these vectors are marked in green and enclose the green area. This again has the same area (or volume) as the blue and pink regions and denotes the third Brillouin zone.

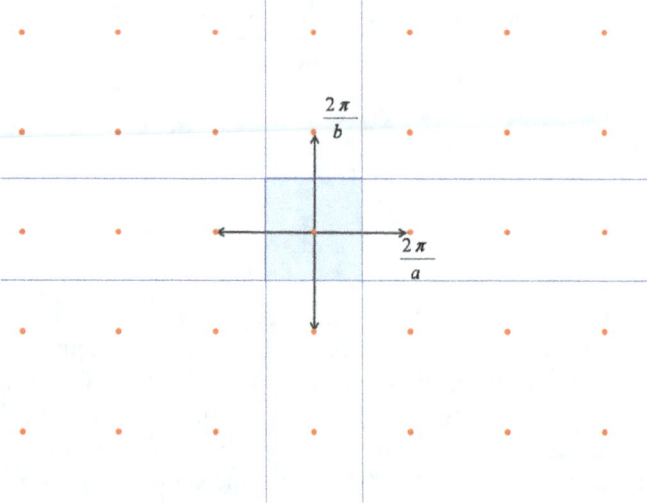

Figure 5.9. Construction of first Brillouin zone (blue) for a square reciprocal lattice.

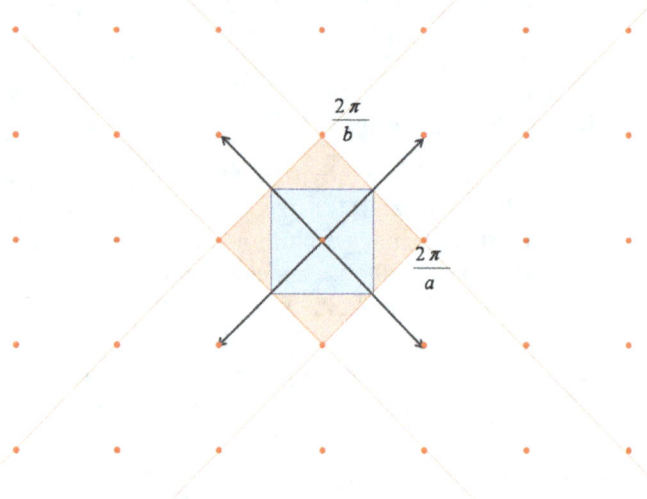

Figure 5.10. Construction of second Brillouin zone (pink) for a square reciprocal lattice.

Figure 5.12 shows the higher-order zones for a square lattice. It is obvious that in principle one could continue with this process, but you can see that the diagrams become ever more complicated. For lower-symmetry reciprocal lattices the complexity of doing this is hardly worth the effort. As a result, in practice, wave information is usually not presented in this form but, instead, use is made of the so-called *reduced zone scheme*.

Definition. *The pieces of each higher zone can be transferred into the first zone through moving the separate regions by subtracting appropriate reciprocal lattice*

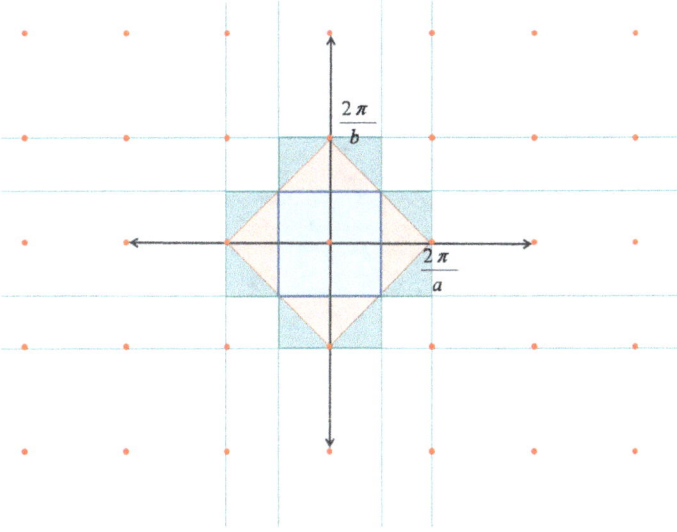

Figure 5.11. Construction of third Brillouin zone (green) for a square reciprocal lattice.

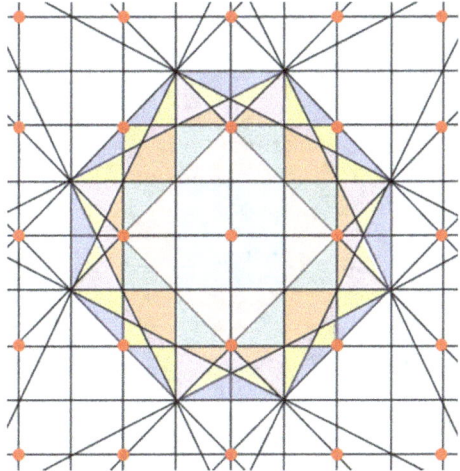

Figure 5.12. Wigner–Seitz zones for a two-dimensional square lattice (first 7 zones shown coloured) [3].

vectors, and the states of each zone form a continuous energy band over the first zone; this is called the reduced zone scheme, whereas use of all Brillouin zones (first, second, third, etc) is known as the extended zone scheme.

5.7 Density of states

We now turn to a most important topic in connection with describing the wave-states and how they relate to energy or frequency. The concept of the density of

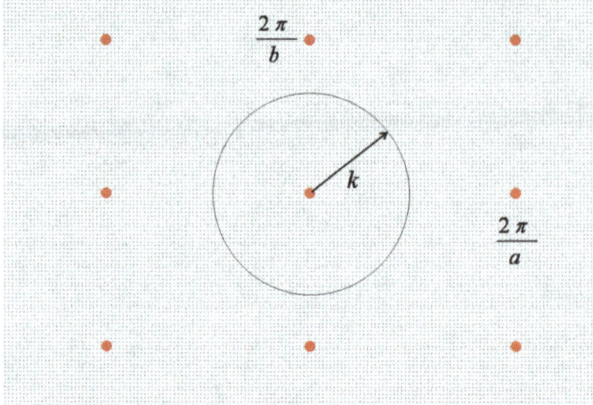

Figure 5.13. Calculation of number of states within a sphere of radius k.

states lies at the heart of much of statistical mechanics and therefore is important for understanding behaviour such as the thermal and electronic properties of a crystalline solid.

We start by working out the number of states within a wave-vector range. We shall do this for an isotropic system, as this is sufficient for most purposes.

We start by calculating the number of wave-states within a bounded region in **k**-space. In three dimensions, this bounded region is a sphere. Figure 5.13 shows a sphere of radius k containing a number of states N_{states}. Now each state is separated from its neighbour by $2\pi/L$, and so the volume around each state is given by

$$V_{\text{state}} = \left(\frac{2\pi}{L}\right)^3 = \frac{8\pi^3}{V_{\text{crystal}}} \tag{5.24}$$

N_{states} is then found by dividing the volume of the sphere by V_{crystal}:

$$N_{\text{states}} = \frac{4}{3}\pi k^3 \div \frac{8\pi^3}{V_{\text{crystal}}} = \frac{V_{\text{crystal}}k^3}{6\pi^2} \tag{5.25}$$

Then, because in reality the number of states is enormous for a normal-sized crystal, we can treat N_{states} as representing a continuum, and so we can calculate the number of occupied states within a range dk by

$$\frac{dN_{\text{states}}}{dk} = \frac{V_{\text{crystal}}k^2}{2\pi^2} \tag{5.26}$$

This defines the *density of states* in k-space.

Definition. *The density of states in k-space is the number of states within the range k and $k + dk$.*

Table 5.2. Numbers and density of k states for different dimensionalities.

Dimension	N_{states}	dN_{states}/dk
1	$\dfrac{L_{crystal}k}{2\pi}$	$\dfrac{L_{crystal}}{2\pi}$
2	$\dfrac{A_{crystal}k^2}{4\pi}$	$\dfrac{A_{crystal}k}{2\pi}$
3	$\dfrac{V_{crystal}k^3}{6\pi^2}$	$\dfrac{V_{crystal}k^2}{2\pi^2}$

This, however, is not the usual way in which this concept is used. Instead it is the energy (or because $E = \hbar\omega$ frequency) density of states that is employed. To find this, we make use of the identity

$$D(E) = \frac{dN_{states}}{dE} = \frac{dN_{states}}{dk}\frac{dk}{dE} \qquad (5.27)$$

or

$$D(\omega) = \frac{dN_{states}}{d\omega} = \frac{dN_{states}}{dk}\frac{dk}{d\omega} \qquad (5.28)$$

We therefore modify our definition of the density of states as follows.

Definition. *The energy (or frequency) density of states $D(E)$ or $D(\omega)$ is the number of states within the range E and $E + dE$ (or ω and $\omega + d\omega$).*

In order to proceed further we need to evaluate either dk/dE or $dk/d\omega$, and so we need a formula relating E (or ω) to k. This depends on the specific problem that needs addressing. We shall see later how this is used in understanding the thermal and electronic properties of crystalline solids.

The formulae (5.25) and (5.26) were found for a three-dimensional system. Clearly this can be generalized for any dimensional system. Some examples are given in table 5.2.

Notice that the density of states is proportional to the size of the crystal, so that for a large crystal the number of wave-states contained within the Brillouin zone is huge (and so can be treated as a continuum of states). As the crystal size diminishes, the distances between the wave-vectors for the states increases thus lowering the number of wave-states within the Brillouin zone.

References

[1] Bradley C J and Cracknell A P 1972 *The mathematical theory of symmetry in solids: representation theory for point groups and space groups* (Oxford: Oxford University Press)

[2] Glazer A M and Burns G 2013 *Space Groups for Solid State Scientists* (Oxford: Oxford University Press)

[3] Brillouin L 1946 *Wave Propagation in Periodic Structure.* (New York: McGraw-Hill)

IOP Concise Physics

A Journey into Reciprocal Space
A crystallographer's perspective
A M Glazer

Chapter 6

Thermal and electronic properties

If you can't stand the heat, then get out of the kitchen!

Harry S Truman

6.1 Specific heat capacity of solids

In this final chapter, we shall look at a couple of physical properties of solids that illustrate the principles described so far, starting with thermal effects. There has long been a historical interest in determining the heat capacity of solids, going back to 1819, when Pierre Louis Dulong and Alexis Therese Petit found, or so they thought, that all solids had the same value for the heat capacity multiplied by the atomic weights. The Law of Dulong and Petit can be related to Maxwell's equipartition theorem whereby the internal energy is given by ½ $k_B T$ per degree of freedom. k_B is Boltzmann's constant.

Definition. *The number of degrees of freedom of a system is the number of independent variables needed to describe that system.*

The thermal behaviour of a solid is determined by the vibrations of its atoms. Now, a vibrational mode has *two* degrees of freedom, one for potential energy and one for kinetic energy[1]. According to Maxwell, in a three-dimensional solid each atom should provide an energy contribution of $2 \times 3k_B T/2$ for vibration in all three dimensions, i.e., three degrees of freedom × 2. Therefore, for a solid the total internal energy is given by

$$E = 3N_A k_B T. \tag{6.1}$$

---

[1] It is a common mistake to equate the number of degrees of freedom to the number of normal modes of a system. While this is true for molecular translations and rotations, in the case of vibrational modes there is a factor of two for the number of degrees of freedom.

where N_A is Avogadro's number: $6.022\,141\,5 \times 10^{23}$ mol^{-1}. Then the specific heat capacity at constant volume is

$$C_V = 3N_A k_B = 3R. \tag{6.2}$$

where R is the gas constant. This is equal to the Dulong and Petit value of 24.94 J K^{-1} mole^{-1}. Nice though this is, it turned out that this is only true at high temperatures, and at lower temperatures the specific heat capacity is lower and temperature-dependent. As a result, several theories were created to explain the observed behaviour, the two most popular being those due to Einstein and to Debye.

6.2 Einstein model

Einstein's approach owed much to the work of others such as Planck and Boltzmann, and is in fact a simple modification of the Dulong and Petit model. In order to introduce a temperature-dependence into the theory he considered the atomic vibrations to be quantized harmonic oscillators all with a single frequency ω_E. This is obviously a gross approximation but nonetheless, as we shall see, it does have features that fit rather well with the observed data. For a quantized harmonic oscillator

$$E = \left(n + \frac{1}{2}\right)\hbar\omega. \tag{6.3}$$

The single-particle partition function is given by

$$\begin{aligned}
Z_{sp} &= \sum_n e^{-\left(n+\frac{1}{2}\right)\hbar\omega_E/k_B T} \\
&= e^{-\frac{1}{2}\hbar\omega_E/k_B T}\left(1 + e^{-\hbar\omega_E/k_B T} + e^{-2\hbar\omega_E/k_B T} + e^{-3\hbar\omega_E/k_B T} + \ldots\ldots\right) \\
&= \frac{e^{-\frac{1}{2}\hbar\omega_E/k_B T}}{1 - e^{-\hbar\omega_E/k_B T}}.
\end{aligned} \tag{6.4}$$

The mean internal energy is then given by

$$\begin{aligned}
E &= 3N_A k_B T^2 \frac{\partial \ln Z_{sp}}{\partial T} \\
&= 3N_A \hbar\omega_E \left\{\frac{1}{2} + \frac{1}{e^{\hbar\omega_E/k_B T} - 1}\right\}.
\end{aligned} \tag{6.5}$$

and then the specific heat capacity is

$$\begin{aligned}
C_V &= \frac{3N_A}{k_B T^2} \frac{\hbar\omega_E e^{\hbar\omega_E/k_B T}}{(e^{\hbar\omega_E/k_B T} - 1)^2} \\
&= 3R\left(\frac{\Theta_E}{T}\right)^2 \frac{e^{\Theta_E/T}}{(e^{\Theta_E/T} - 1)^2}.
\end{aligned} \tag{6.6}$$

and the *Einstein temperature* $\Theta_E = \hbar\omega_E/k_B$. For high temperatures, where $T \gg \Theta_E$, $k_B T \gg \hbar\omega_E$ and this expression tends to

$$C_V \to 3R. \qquad (6.7)$$

the Dulong and Petit law. On the other hand, for very low temperatures

$$C_V \to 3R\left(\frac{\Theta_E}{T}\right)^2 e^{-\Theta_E/T}. \qquad (6.8)$$

which tends to zero as $T \to 0$.

Figure 6.1 shows the agreement between the Einstein model and observed specific heat capacity measurements for several metals, where it can be seen that, despite the crudeness of the approximation, the model stands up rather well. In reality, however, the specific heats of solids do not approach zero quite as quickly as suggested by Einstein's model when $T \to 0$. In practice, it looks more like

$$C_V \propto T^3. \qquad (6.9)$$

(figure 6.2) rather than the more complicated exponential behaviour of equation (6.6).

6.3 Debye model

A more complete description of the specific heat capacity was achieved by Peter Debye in 1912. Where Einstein's model can be considered to be based on oscillating

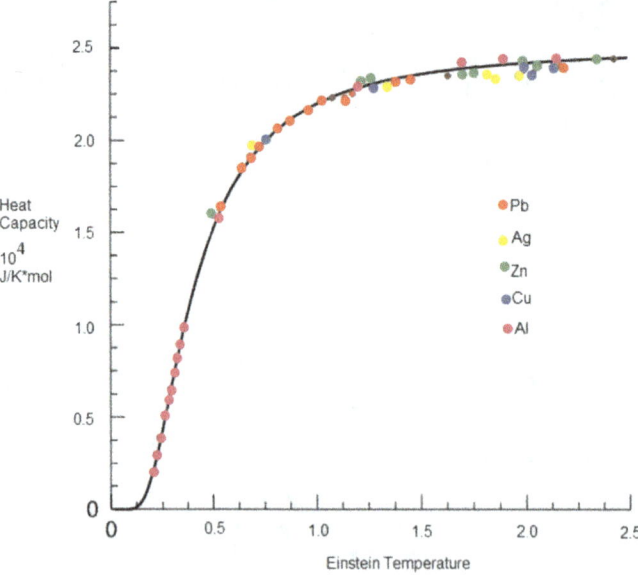

Figure 6.1. The Einstein specific heat capacity as a function of temperature (normalized to the Einstein temperature) for several metals [1].

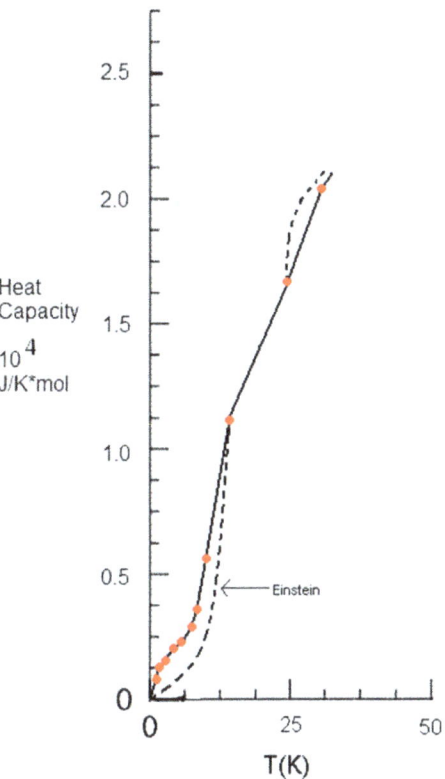

Figure 6.2. The Einstein specific heat capacity as a function of temperature (normalized to the Einstein temperature) at low temperature [1].

atoms, Debye's treatment is at the other extreme in which case the solid is treated rather like an elastic continuum and atoms are not considered *per se*. Here, the thermal energy is involved in creating waves that distort the solid in some way. Using the language of quantum mechanics, then, such waves can also be interpreted in terms of particles, which in this case are called *phonons*. The name is derived from the fact that typical low-frequency waves are in many respects similar to sound waves. Phonons are described by symmetric wave functions and so, from a statistical mechanics point of view, they are bosons. As they have no physical existence outside the solid, they are in fact known as *quasiparticles*.

Debye makes the following assumptions:
1 The solid can be treated as if it were an elastic continuum.
2 Phonons are created as temperature increases.
3 There is no frequency dispersion in the solid, i.e., $\omega = vk$, where v is the wave velocity (figure 6.3(a)).
4 There is a cut-off frequency ω_D (and wave-vector k_D) above which no phonons exist in the solid (figure 6.3(a)).

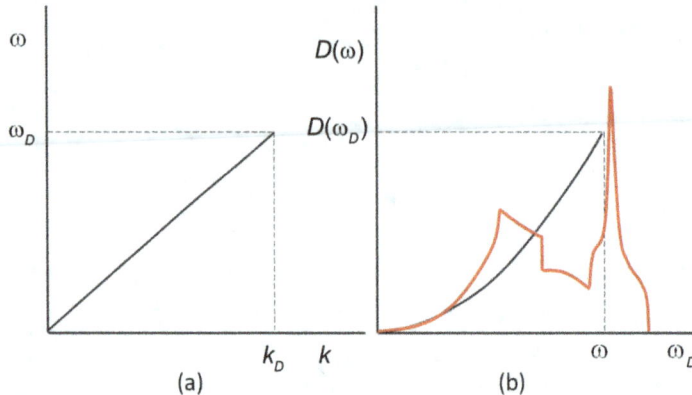

Figure 6.3. Assumptions of the Debye model of specific heat capacities. (a) No dispersion. (b) Parabolic density of states. Red curve experimental trace for aluminium.

For a three-dimensional solid the density of states is

$$D(\omega) = 3 \times \frac{V_{\text{crystal}} k^2}{2\pi^2} \frac{dk}{d\omega}. \tag{6.10}$$

An extra factor of three has been added to account for the fact that the phonons have three polarizations, two of which are transverse and one of which is longitudinal. Since there is no dispersion in this approximation

$$D(\omega) = \frac{3 V_{\text{crystal}} \omega^2}{2\pi^2 v^3}. \tag{6.11}$$

This is shown in figure 6.3(b). The number of states is then given by

$$\begin{aligned} N_{\text{states}} &= \frac{3 V_{\text{crystal}}}{2\pi^2 v^3} \int_0^{\omega_D} \omega^2 \, d\omega \\ &= \frac{V_{\text{crystal}}}{2\pi^2 v^3} \left(\frac{k_B \Theta_D}{\hbar} \right)^3. \end{aligned} \tag{6.12}$$

where Θ_D is the so-called Debye temperature. The integral is used here since we assume the sample to be sufficiently large that the density of states forms a quasicontinuum. The upper limit of the integral is set at the Debye cut-off frequency. Since we are dealing with phonons, we can now introduce the Bose–Einstein factor $f(\omega, T)$ but with the chemical potential $\mu = 0$, as the numbers of phonons are not conserved as temperature changes:

$$f(\omega, T) = \frac{1}{e^{\hbar \omega / k_B T} - 1}. \tag{6.13}$$

The internal energy is then given by

$$\begin{aligned}E &= \int_0^{\omega_D} \hbar\omega D(\omega) f(\omega, T) d\omega \\ &= \frac{3V_{\text{crystal}}}{2\pi^2 v^3} \int_0^{\omega_D} \frac{\hbar\omega^3}{e^{\hbar\omega/k_B T} - 1} d\omega \\ &= 3N_{\text{states}} \left(\frac{\hbar}{k_B \Theta_D}\right)^3 \int_0^{\omega_D} \frac{\hbar\omega^3}{e^{\hbar\omega/k_B T} - 1} d\omega.\end{aligned} \qquad (6.14)$$

This integral is fearsome and in order to simplify it we make the substitution

$$x = \frac{\hbar\omega}{k_B T}. \qquad (6.15)$$

and then

$$E = 3N_{\text{states}} k_B T \left(\frac{T}{\Theta_D}\right)^3 \int_0^{x_D} \frac{x^3}{e^x - 1} dx. \qquad (6.16)$$

Now, for high temperatures $k_B T \gg \hbar\omega$ the integral is close to zero. We can then make the following approximation

$$\frac{x^3}{e^x - 1} \simeq \frac{x^3}{1 + x - 1} = x^2. \qquad (6.17)$$

and then

$$\begin{aligned}E &= 3N_{\text{states}} k_B T \left(\frac{T}{\Theta_D}\right)^3 \int_0^{x_D} x^2 \, dx \\ &= N_{\text{states}} k_B T \left(\frac{T}{\Theta_D}\right)^3 [x^3]_0^{x_D}.\end{aligned} \qquad (6.18)$$

But

$$x_D = \frac{\Theta_D}{T}. \qquad (6.19)$$

and so

$$\begin{aligned}E &= 3N_{\text{states}} k_B T \\ \therefore C_V &= 3N_{\text{states}} k_B.\end{aligned} \qquad (6.20)$$

I need to inject a caution here. The factor N_{states} appears in this result rather than N_A, Avogadro's number. It is a common mistake in textbooks to confuse numbers of states with numbers of atoms. In this case, if we can assume that the number of states equals the number of atoms in the crystal then we get

$$C_V = 3N_A k_B. \tag{6.21}$$

the Dulong and Petit law. However, one has to be careful here because you cannot generally assume that the number of states is equal to the number of atoms. The number of states has nothing to do with number of atoms but it is equal to the number of primitive unit cells in the crystal (note that our derivation of density of states does not involve atoms at all!). If we write $N_{\text{states}} = N_A$ we are, in fact, assuming that there is one atom per primitive unit cell. So why do we get away with this formulation? The reason is subtle and is due to the type of mode of vibration that is primarily responsible for heat capacity. As we shall see later, every atom can oscillate in three orthogonal directions, so that if in any primitive unit cell there are N atoms there will be $3N$ modes of vibration. The three lowest-energy modes (the so-called acoustic modes) are the main contributors to the Debye heat capacity because their frequencies are relatively low and correspond to thermal frequencies, while the remaining $3N - 3$ modes are generally of much higher frequency (the optic modes) and so play a much smaller role. As every crystal has at least one atom in the primitive unit cell the heat capacity is dominated by the three acoustic modes, and this is why in practice we can treat the number of states in the Debye approximation as being equivalent to having one atom in the primitive unit cell.

Now consider what happens at low temperatures where $k_B T \ll \hbar \omega$. In this case, we are well below the Debye temperature and so the integral can be taken from 0 to ∞.

$$E = 3N_{\text{states}} k_B T \left(\frac{T}{\Theta_D}\right)^3 \int_0^\infty \frac{x^3}{e^x - 1} dx. \tag{6.22}$$

The integral can be evaluated as follows

$$\begin{aligned}
\int_0^\infty \frac{x^3}{e^x - 1} dx &= \int_0^\infty \left(\sum_{n=1}^\infty x^3 e^{-nx}\right) dx \\
&= \sum_{n=1}^\infty \left(\int_0^\infty x^3 e^{-nx} dx\right) \\
&= \sum_{n=1}^\infty I_n.
\end{aligned} \tag{6.23}$$

Integrating by parts we get

$$
\begin{aligned}
I_n &= \int_0^\infty x^3 e^{-nx}\, dx \\
&= \left[-x^3 \frac{e^{-nx}}{n}\right]_0^\infty + \int_0^\infty 3x^2 \frac{e^{-nx}}{n} dx \\
&= 0 + \frac{3}{n}\int_0^\infty x^2 \frac{e^{-nx}}{n} dx \\
&= \frac{3}{n}\left[-x^2 \frac{e^{-nx}}{n}\right]_0^\infty + \frac{3}{n}\int_0^\infty 2x \frac{e^{-nx}}{n} dx \\
&= 0 + \frac{6}{n^2}\int_0^\infty x \frac{e^{-nx}}{n} dx \\
&= \frac{6}{n^2}\left[-x \frac{e^{-nx}}{n}\right]_0^\infty + \frac{6}{n^2}\int_0^\infty \frac{e^{-nx}}{n} dx \\
&= 0 + \frac{6}{n^3}\int_0^\infty \frac{e^{-nx}}{n} dx \\
&= \frac{6}{n^3}\left[-\frac{e^{-nx}}{n}\right]_0^\infty \\
&= \frac{6}{n^4}.
\end{aligned}
\qquad (6.24)
$$

Therefore,

$$
\int_0^\infty \frac{x^3}{e^x - 1} dx = 6\sum_{n=1}^\infty \frac{1}{n^4} = \frac{6\pi^4}{90}. \qquad (6.25)
$$

using the Riemann Zeta function evaluated at $\zeta(4)$. Thus,

$$
\begin{aligned}
E &= 3N_{\text{states}} k_B T \left(\frac{T}{\Theta_D}\right)^3 \frac{6\pi^4}{90} \\
&= \frac{3\pi^4}{5} N_{\text{states}} k_B T \left(\frac{T}{\Theta_D}\right)^3.
\end{aligned}
\qquad (6.26)
$$

The specific heat capacity is then given by

$$
C_V = \frac{12\pi^4}{5} N_{\text{states}} k_B \left(\frac{T}{\Theta_D}\right)^3. \qquad (6.27)
$$

correctly showing that $C_V \propto T^3$ in line with experiment.

The Debye model can be improved in various ways. For instance, instead of assuming that all three acoustic waves have the same velocity, one can use

$$
\frac{1}{v^3} = \frac{1}{v_{\text{trans}}^3} + \frac{1}{v_{\text{trans}}^3} + \frac{1}{v_{\text{long}}^3}. \qquad (6.28)
$$

for the two transverse polarizations and one longitudinal polarization. Note too that the result depends on the dimensionality of the system, since the density of states depends on the dimensionality m of the system. Therefore

$$C_V \propto T^m. \tag{6.29}$$

For instance, crystals of graphite have layers of carbon atoms strongly bound together in planes with weak van der Waals interactions between the layers. Measurements of the specific heat capacity find $C_V \propto T^{2.4}$ at very low temperatures.

The significance of the Debye temperature is that this is the temperature where the maximum number of phonons are created by thermal energy, above which the classical Dulong and Petit behaviour is expected. This is a function of the strength of the bonding between the atoms in the crystal, with strong bonds meaning stiff force constants between the atoms leading to higher-frequency phonons; hence higher Debye temperatures. Thus, for example, crystals of diamond, one of the hardest materials known, and in which carbon atoms are strongly covalently bonded, has a Debye temperature of 2230 K. On the other hand, a softer material such as aluminium has a Debye temperature of 428 K.

Figure 6.4 shows a comparison between the Einstein and Debye models, where it is seen that while both saturate at high temperatures to the same value, the Einstein model underestimates the specific heat capacity increasingly as temperature is decreased.

6.4 Vibrations of atoms

When a crystal is heated the main result is an increase in the vibrational amplitudes and frequencies of the atoms. Because the atoms interact with each other through

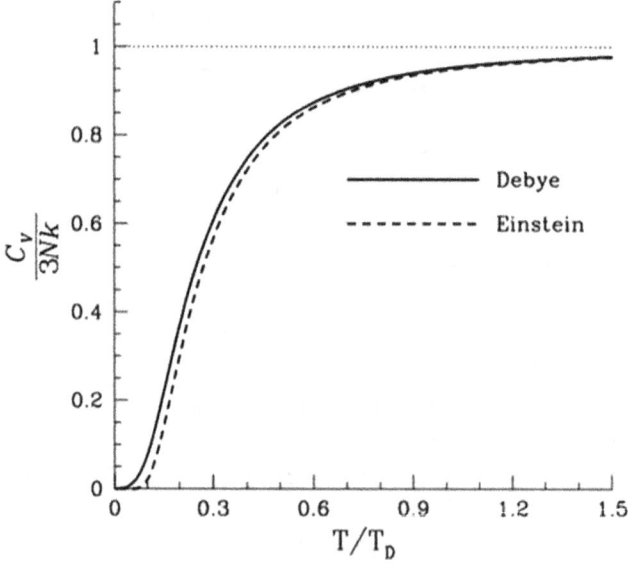

Figure 6.4. Comparison between Debye and Einstein models. From [2].

the periodic potential the motion of one atom has an effect on the surrounding atoms, the result being that wave-like vibrational modes are set up. There are several ways in which this can be described, and we shall see that the concept of reciprocal space plays an integral part in this. We shall begin here with a very simple way of viewing vibrational behaviour in real space from an atomistic point of view using purely classical physics (although we should bear in mind that, in reality, the solutions that we obtain are actually quantum states: we get away with the classical approach because the density of states in a typical solid is very large and the quantum states constitute a quasi-continuum).

One-dimensional monatomic chain

A standard problem, discussed in most solid-state texts, is that of a one-dimensional linear chain of atoms separated by a distance a along which waves can be propagated. In this case the distance a is also the primitive unit cell repeat length. These waves can be transverse (in which case there are two such transverse, degenerate components) or longitudinal (figure 6.5).

To determine the characteristics of this simple system we need to define the classical equation of motion for the atoms. In its simplest form this is done by use of Hooke's Law and thinking of the bonds between the atoms as if they were springs with force constant C. As one atom moves, it influences its neighbours, next-neighbours and so on causing them to move in sympathy. It is this that creates the wave. Here we shall take the trivial approach of just having nearest-neighbour force constants, but it is relatively easy to incorporate higher-order forces if required.

Figure 6.6(a) shows a row of lattice points of spacing a and (b) a set of identical atoms of mass M separated joined by springs. Consider the displacement of the sth

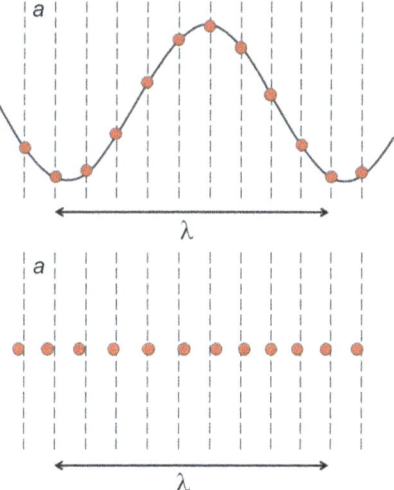

Figure 6.5. Examples of waves propagating on a one-dimensional linear chain of atoms. Upper wave is transverse and lower wave is longitudinal.

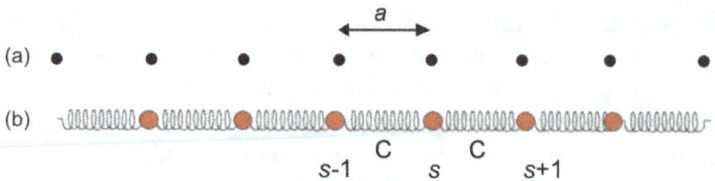

Figure 6.6. A linear chain of atoms of mass M with repeat distance a.

atom denoted by u_s. According to Hooke's Law, the equation of motion is then given by

$$M\ddot{u}_s = C(u_{s+1} - u_s) + C(u_{s-1} - u_s). \tag{6.30}$$

If we now assume a nearest-neighbour wave-like solution of the form

$$u_s = Ue^{i(ka-\omega t)}. \tag{6.31}$$

we then get

$$\begin{aligned}-M\omega^2 &= C(e^{ika} + e^{-ika} - 2) \\ &= 2C(\cos ka - 1).\end{aligned} \tag{6.32}$$

Therefore, we find that the frequency is

$$\omega = \sqrt{\frac{4C}{M}} \left|\sin\frac{ka}{2}\right| = \omega_{max}\left|\sin\frac{ka}{2}\right|. \tag{6.33}$$

The group velocity of this wave is given by

$$v_g = \frac{d\omega}{dk} = \sqrt{\frac{Ca^2}{M}} \cos\frac{ka}{2} = \frac{a\omega_{max}}{2}\cos\frac{ka}{2}. \tag{6.34}$$

The frequency is plotted as a function of wave-vector in figure 6.7 and the group velocity in figure 6.8.

Definition. *The plot of frequency against wave-vector for a vibrating system is known as a dispersion curve.*

The first thing to notice is that both curves are periodic in k-space with a periodicity of $2\pi/a$. Also shown in both figures are the 1st, 2nd and 3rd Brillouin zones (coloured blue, pink and green, respectively) where it can be seen that the same solutions appear in each. It is obvious therefore that all the information needed to construct these curves is contained in one half of the 1st Brillouin zone within the range 0 to π/a. In other words, it is not necessary to consider higher-order Brillouin zones in practice.

Note too that the group velocity is highest at $k = 0$ but drops to zero at the values $\pm(2n + 1)\pi/a$, or for the 1st Brillouin zone at $\pm\pi/a$. Now these points differ by a reciprocal lattice vector $2\pi/a$ and so the solution at π/a is for a travelling wave to the

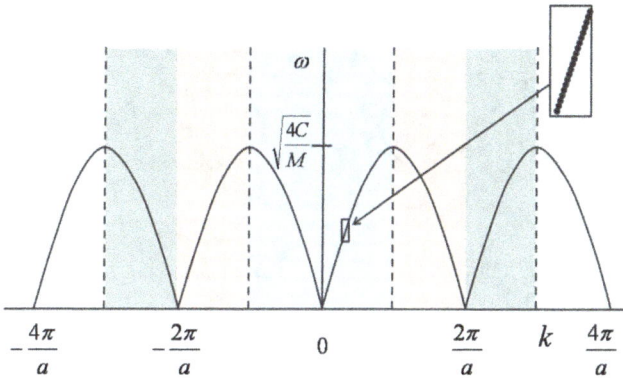

Figure 6.7. Frequency of vibration for a monatomic chain plotted as a function of wave-vector. 1st Brillouin zone = blue: 2nd Brillouin zone = pink: 3rd Brillouin zone = green. The inset diagram is a magnified view of the curve showing that in fact it consists of a number of wave-states close together to form a quasi-continuum. The smaller the length of the chain the fewer such wave-states will exist within the Brillouin zone and the smaller the density of states.

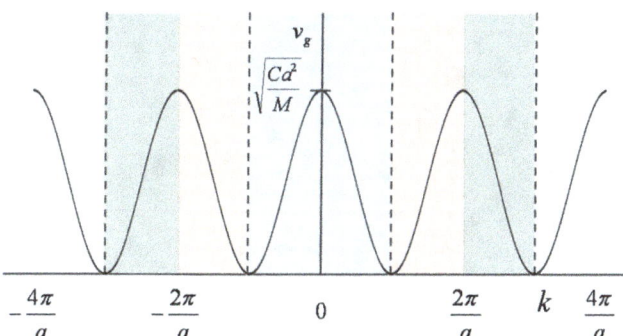

Figure 6.8. Group velocity for monatomic chain as a function of wave-vector. 1st Brillouin zone = blue: 2nd Brillouin zone = pink: 3rd Brillouin zone = green.

right while $-\pi/a$ is for an equal amplitude wave travelling to the left. In other words on the Brillouin zone boundary the solution is that of a standing wave, i.e., two equivalent waves travelling in opposite directions, thus corresponding in total to a standing wave. One can think of this as a wave travelling to the right being reflected back (Bragg scattered) to the left, provided that its wavelength corresponds to a wave-vector ending on the Brillouin zone boundary.

It can be seen therefore that there are in principle for any wave an infinite number of solutions consistent with the periodicity of the reciprocal and real lattices. In general, the wave-vector solutions are given by $\pm k \pm \frac{2n\pi}{a}$ where n is an integer. Figure 6.9 shows three such wave-vector solutions corresponding to $k = \frac{\pi}{2a}, \frac{2\pi}{a} - k$ and $\frac{4\pi}{a} - k$, the second and third solutions corresponding to waves travelling in the opposite direction to the first. All these solutions are equivalent, but it can also be seen that only the wave with $k = \frac{\pi}{2a}$ need be considered, as all other solutions carry no further

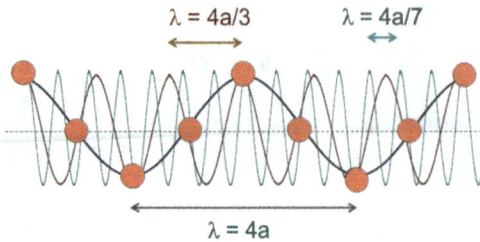

Figure 6.9. Example of three transverse waves for the monatomic chain. Although the wavelengths are different they are all possible solutions.

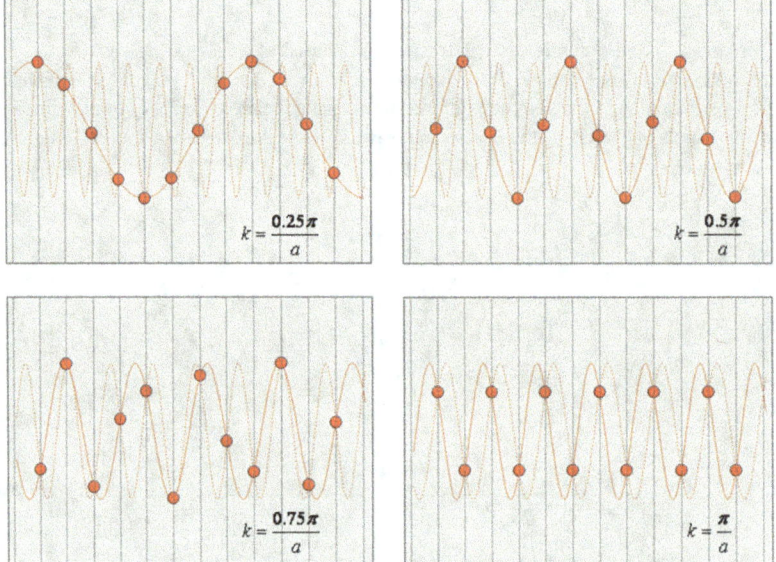

Figure 6.10. Examples of transverse oscillations in a monatomic chain for different values of wave-vector Simulations by the program CHAINPLOT obtainable from [3].

information. I tend to think of these other solutions as being effectively imaginary or unphysical. The wave with $k = \frac{\pi}{2a}$ is the only one that lies *within* the 1st Brillouin zone.

Therefore, in general, starting from the centre of the 1st Brillouin zone, where $k = 0$ and $\lambda = \infty$, as k increases the wavelength decreases until the point where k reaches the Brillouin zone boundary. Here $k = \pi/a$ and the wavelength is equal to $2a$. All wave-vectors outside the Brillouin zone boundary then have shorter wavelengths, which we can regard as being unphysical. Therefore, the wavelengths must be $\geqslant 2a$ to be considered as physically useful solutions.

Figure 6.10 shows some examples of transverse waves computed by the program CHAINPLOT [3]. The dashed curves show the solutions where a reciprocal lattice vector has been subtracted, corresponding to an unphysical wave travelling to the left. Only for $k = \pi/a$ (diagram at bottom right) do we obtain two physically meaningful solutions travelling in opposite directions to form a standing wave.

Finally, consider the density of states for this system. For a one-dimensional system this is given by

$$D(\omega) = \frac{3L}{\pi} \frac{dk}{d\omega}. \qquad (6.35)$$

The factor of three arises here because there are three possible modes, two transverse and one longitudinal.

Therefore,

$$\begin{aligned}
\frac{d\omega}{dk} &= \frac{a\omega_{max}}{2} \cos \frac{ka}{2} \\
&= \frac{a\omega_{max}}{2} \left(1 - \sin^2 \frac{ka}{2}\right)^{\frac{1}{2}} \\
&= \frac{a\omega_{max}}{2} \left(1 - \left[\frac{\omega}{\omega_{max}}\right]^2\right)^{\frac{1}{2}} \\
&= \frac{a}{2} \left(\omega_{max}^2 - \omega^2\right)^{\frac{1}{2}}.
\end{aligned} \qquad (6.36)$$

and this results in

$$D(\omega) = \frac{6L}{\pi a} \frac{1}{\left(\omega_{max}^2 - \omega^2\right)^{\frac{1}{2}}}. \qquad (6.37)$$

The density of states, therefore, shows a singularity when $\omega = \omega_{max}$ (corresponding to $k = \pm\pi/a$, the zone boundary value).

One-dimensional diatomic chain

We now consider what happens if within the repeat distance a there are two atoms of differing mass M_1 and M_2. Figure 6.11 shows such an alternating array. It is most important to emphasize that the distance a, as indicated for the lattice in figure 6.11(a), is not the distance between atoms but it is in fact the *repeat* distance. In other words, it is a unit cell length for a one-dimensional primitive unit cell containing two different atoms.

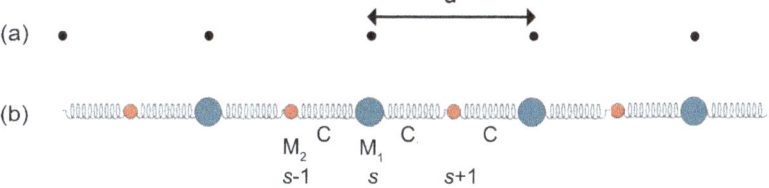

Figure 6.11. The diatomic chain with alternating atoms of different mass $M_1 > M_2$ but with equal force constants C between them.

Once again using Hooke's Law we now have to consider two equations of motion:
$$M_1\ddot{u}_s = C[u_{s+1} - u_s] + C[u_{s-1} - u_s]$$
$$M_2\ddot{u}_{s+1} = C[u_{s+2} - u_{s+1}] + C[u_s - u_{s+1}].$$
(6.38)

Then assuming two wave-like solutions
$$u_s = Ue^{i(ka/2-\omega t)}$$
$$u_{s+1} = Ve^{i(ka/2-\omega t)}.$$
(6.39)

This gives rise to the following two equations
$$-M_1\omega^2 U = 2CV\cos k\frac{a}{2} - 2CU$$
$$-M_2\omega^2 V = 2CU\cos k\frac{a}{2} - 2CV.$$
(6.40)

For a non-trivial solution
$$\begin{vmatrix} 2C - M_1\omega^2 & 2C\cos k\frac{a}{2} \\ 2C\cos k\frac{a}{2} & 2C - M_2\omega^2 \end{vmatrix} = 0.$$
(6.41)

Hence
$$(2C - M_1\omega^2)(2C - M_2\omega^2) - 4C^2\cos^2\frac{ka}{2} = 0.$$
(6.42)

Finally, this gives the result
$$\omega^2 = \frac{C(M_1 + M_2)}{M_1 M_2} \pm C\left[\left(\frac{M_1 + M_2}{M_1 M_2}\right)^2 - \frac{4}{M_1 M_2}\sin^2 ka/2\right]^{\frac{1}{2}}.$$
(6.43)

This is plotted in figure 6.12. As with the monatomic chain we see that the dispersion is periodic in the reciprocal lattice with Brillouin zone boundaries at $\pm(2n + 1)\pi/a$. However, this time for any value of wave-vector **k** there are two solutions, one at low frequency and one at high frequency. The lower-frequency curve is known as the acoustic branch and the higher-frequency curve is called the optic branch. The acoustic branch is so-called because at small values of **k** the frequency is approximately linearly related to the wave-vector, i.e., it is almost dispersionless, and in real solids corresponds to a velocity similar to that for sound waves. The optic branch, on the other hand, corresponds to frequencies typically seen in the optical region. The amplitudes U and V are given by

$$\frac{U}{V} = \frac{2C\cos k\frac{a}{2}}{2C - M_1\omega^2} \quad \text{acoustic branch}$$
$$\frac{V}{U} = \frac{2C\cos k\frac{a}{2}}{2C - M_2\omega^2} \quad \text{optic branch.}$$
(6.44)

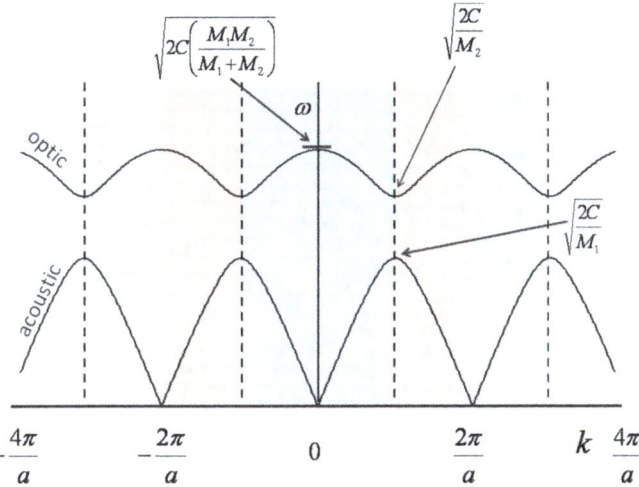

Figure 6.12. Dispersion curves for a diatomic chain in which $M_1 > M_2$. 1st Brillouin zone = blue; 2nd Brillouin zone = pink; 3rd Brillouin zone = green.

Notice how on the Brillouin zone boundaries, where $k = \pm\pi/a$, the two solutions are

$$\omega_{\text{acoustic}} = \sqrt{\frac{2C}{M_1}} \text{ and } U = 0$$

$$\omega_{\text{optic}} = \sqrt{\frac{2C}{M_2}} \text{ and } V = 0. \tag{6.45}$$

This tells us that, for this value of wave-vector, the acoustic mode only involves motion of the heavy atoms and the light ones are not oscillating at all. On the optic branch at this wave-vector it is the lighter atoms that are vibrating while the heavier ones remain stationary. Clearly if we make the two masses become closer in magnitude the frequency gap between optic and acoustic modes will become narrower. It is therefore instructive to consider what happens to the dispersion curve as the two atoms become increasingly alike (figure 6.13).

In this diagram are plotted the 1st and 2nd Brillouin zones with the heavy line marking the diatomic dispersion curves. If we now imagine that we somehow magically managed to change the masses until $M_1 = M_2$, the gap at the zone boundary closes up and the dispersion curve follows along the dashed trace. But, of course, when this happens we are simply back to the monatomic case. This means that the Brillouin zone boundary is moved from π/a to $2\pi/a$. In this case the repeat distance now becomes equivalent to the interatomic spacing $a_0 = 2a$, so that the Brillouin zone boundary is now at π/a_0. In this situation, the optic branches no longer exist and only the acoustic branch is left (figure 6.14).

This tells us a couple of important things. First of all, there will always be acoustic branches for vibrations in a solid: in general, there will be three such branches, two transverse and one longitudinal, although they may turn out for symmetry reasons to be frequency-degenerate. Secondly, this also shows that when there is more than

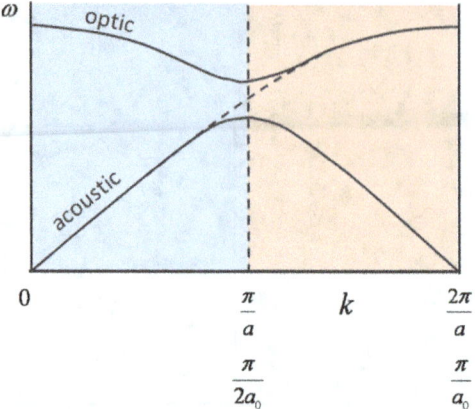

Figure 6.13. Diatomic chain related to monatomic chain.

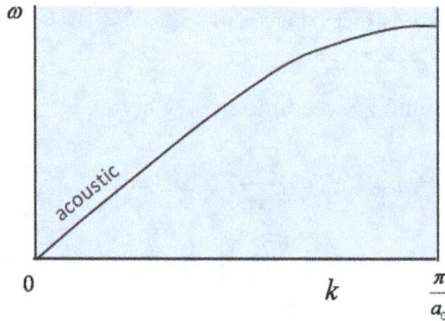

Figure 6.14. Result of making the two atoms in a diatomic chain equivalent in mass. 1st Brillouin zone = blue.

one atom in the primitive unit cell there will be optic branches as well, so that at any value of k we have an energy level diagram with the optic modes as excited states of the acoustic modes (ground states). Unlike in atomic physics, these energy level diagrams depend on the choice of wave-vector.

So now we can begin to understand a bit more about the significance of the extended Brillouin zones. Imagine doing the reverse of the procedure above by starting from the monatomic chain and then causing every second atom to have a higher mass, thus changing the system back to a diatomic chain. Figure 6.15 shows what would happen just as the two types of atom are differentiated. Half-way between the zone centre and the zone boundary for the monatomic case splitting of the curve occurs and this then becomes the new Brillouin zone boundary. This alternative view has the optic branch in the 2nd Brillouin zone and the acoustic branch in the 1st Brillouin zone. So, one way of thinking about this is to say that the ground state terms lie in the 1st Brillouin zone and the first excited states lie in the 2nd Brillouin zone. It is by appealing to reciprocal lattice periodicity that we can fold all the solutions across the Brillouin zone boundaries to obtain figure 6.12, and then instead of using an extended zone scheme use a reduced zone scheme. The reduced zone scheme (figure 6.16) then shows the energy level diagram.

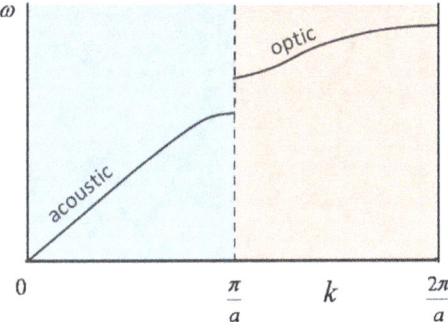

Figure 6.15. Alternative view of the dispersion curves for a diatomic chain.

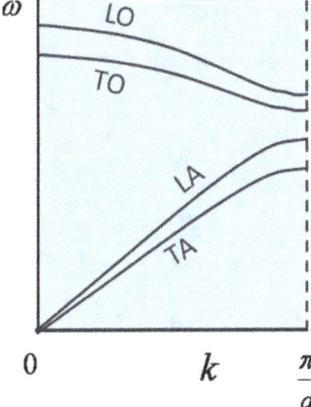

Figure 6.16. Reduced zone scheme for a diatomic chain showing longitudinal acoustic (LA), longitudinal optic (LO), transverse acoustic (TA) and transverse optic (TO) branches.

In this diagram, the longitudinal and transverse acoustic and optic branches are plotted. Bonds are compressed and stretched in longitudinal motion, whereas in transverse motion it is the bond angles that change. The former involves higher force constants and so we expect to find that the longitudinal modes are of higher frequency than the transverse modes. Furthermore, in this example there are two degenerate transverse oscillations with the atomic motions perpendicular to each other, and so we only see single TA and TO curves plotted.

Note. *For any crystal, the number of normal mode branches at any value of k will be equal to 3N where N is the number of atoms in the **primitive** unit cell.*

Figures 6.17 and 6.18 illustrate some simulations of the transverse motion of the atoms in a diatomic chain. The dashed curves represent solutions where a reciprocal lattice vector has been subtracted, and correspond to waves travelling in the opposite direction to those described by the full curves.

We see that, in the acoustic case for $k < \pi/a$, the tendency is for all atoms to move in the same direction, particularly obvious for low values of k. For $k \neq 0$ or π/a, the solutions correspond to running waves and the heavy and light atoms execute

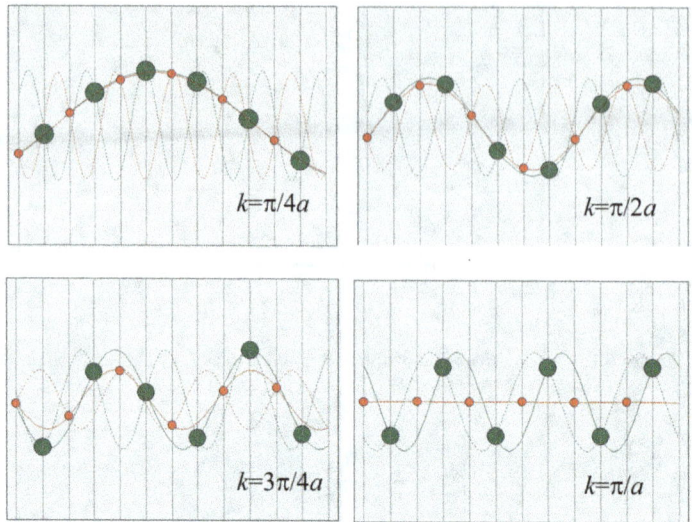

Figure 6.17. Examples of transverse oscillations in a diatomic chain for different values of wave-vector (acoustic branch). Simulations by the program CHAINPLOT obtainable from [3].

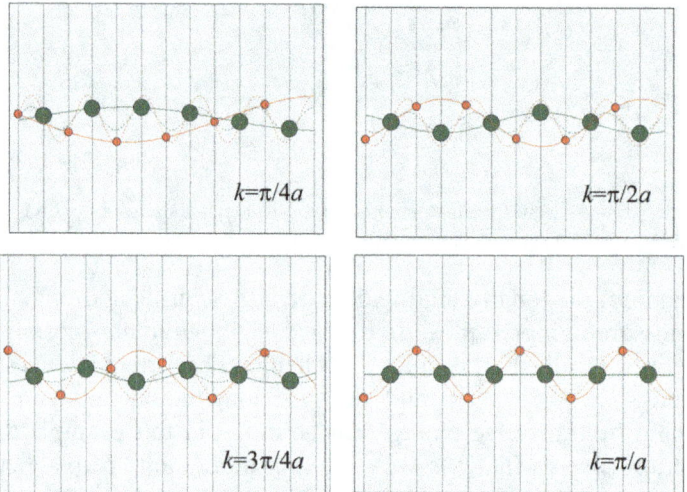

Figure 6.18. Examples of transverse oscillation in a diatomic chain ($M_1/M_2 = 2$) for different values of wave-vector (Optic branch). Simulations by the program CHAINPLOT obtainable from [3].

motion described by two waves, according to equations (6.44) and (6.39) (red and green sinusoidal traces). The dashed traces are for corresponding waves where a reciprocal lattice vector has been subtracted from k and thus correspond to running waves travelling in the opposite direction. As k approaches zero, all the atoms move in the same direction but with very low frequency. Generalizing to a real solid, such low-frequency acoustic modes therefore lead to bulk distortions of the solid, and so their study is important for the understanding of its elastic properties. When $k = \pi/a$, the Brillouin zone boundary, the light atoms remain stationary and the heavy atoms oscillate about their mean positions in accordance with the solutions (6.45). The dashed and full curves in this case constitute a standing wave solution. The effect is to double the unit cell repeat from a to $2a$ (the minimum wavelength for a physical solution). If we were to freeze the motion in this configuration, such as might occur at a phase transition, the result would be that we would have to use a new reciprocal lattice with new lattice points halfway between the former points.

In the optic case (figure 6.18), the tendency is for the two types of atom to act in opposing directions. The frequencies here are high, typically similar to optical frequencies. At very low k the chain of heavy atoms will move in one direction while the light ones will move in the opposite directions. If the two types of atoms happen also to be ions, then this will result in charge separation, i.e., oscillating dipoles (for example in NaCl). The existence of oscillating dipoles will mean that a solid with this type of motion would absorb electromagnetic radiation, typically in the infrared region. Again, when $k = \pi/a$, the motion effectively doubles the unit cell repeat length as in the acoustic case, but this time it is only the light atoms or only the heavy atoms that oscillate. We can also see that a study of low-k optic modes gives information about relative atomic motions within a unit cell.

These simulation diagrams also illustrate something that seems to be incorrect in almost all books. Figure 6.19 shows a couple of diagrams drawn as seen in most textbooks. Notice how the light and heavy atoms seem to sit on a single sinusoidal wave. What has been forgotten here is that, except for $k = \pi/a$, the wave solutions are running waves, not stationary! You might think that because we have used periodic boundary conditions the waves must all be stationary, but remember that dispersion curves, such as those shown in figure 6.16, are usually plotted as travelling waves in half the Brillouin zone, with the other unplotted half corresponding to equivalent waves travelling in the opposite direction. It is when taken together that one has stationary waves. Therefore if, for example, the acoustic case really looked like that shown in figure 6.19, then as the crest of the wave moves from left to right both heavy and light atoms would have to execute the same amplitude. But this cannot be so as equation (6.39) demonstrates. In reality, there are two sinusoidal waves with

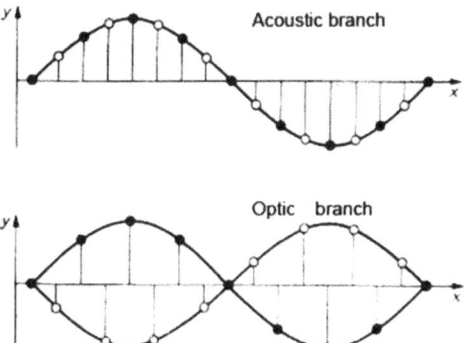

Figure 6.19. Typical diagrams of atomic displacements for the diatomic chains as seen in most textbooks, e.g. Kittel [6].

the same frequency and phase but with different amplitudes, one for the heavy atoms and one for the light atoms, as seen in the simulations earlier. In general, we can say that each atom must lie on its own travelling sine wave. Figure 6.19 is therefore misleading by giving the wrong impression. The diagrams in the books by Brillouin [4] and by Wannier [5] are among the few that are actually correct.

6.5 Lattice dynamics

Real solids are generally not one-dimensional, and so we need a way to handle the more complicated case of a three-dimensional crystal. The term lattice dynamics is used to describe this subject, although this title demonstrates the usual confusion between atoms and lattice points. It is not the lattice that vibrates but the atoms! So, it would be better if we called this atom dynamics, but I guess it is too late for that! The topic is a difficult one and for most cases analytically insoluble, even in the harmonic approximation, but nonetheless, in order to emphasize its complexity, I shall give a brief introduction to the subject here.

Once again, we shall adopt a semi-classical approach. Let us start by defining the kinetic energy of the atoms thus

$$T = \frac{1}{2} \sum_{lsa} M_s \dot{u}_\alpha\binom{l}{s}. \quad (6.46)$$

Here l is the unit cell index, s the atom index within a unit cell and α the Cartesian component of direction. The potential energy can be expressed as a power series via a Taylor expansion in the atomic displacements:

$$V = V_0 + \sum_{ls}\sum_{\alpha} \phi_\alpha\binom{l}{s} u_\alpha\binom{l}{s} + \frac{1}{2} \sum_{\substack{ls \\ l's'}} \sum_{\alpha\beta} \phi_{\alpha\beta}\binom{ll'}{ss'} u_\alpha\binom{l}{s} u_\beta\binom{l'}{s'} + \cdots \quad (6.47)$$

$\phi_{\alpha\beta}\binom{ll'}{ss'}$ represents the force in the α direction on atom ls when the atom $l's'$ is displaced along β. In general, we can write

$$V = V_0 + V_1 + V_2 + \cdots. \quad (6.48)$$

and then

$$\phi_\alpha\binom{l}{s} = \left[\frac{\partial V}{\partial u_\alpha\binom{l}{s}}\right]_0. \quad (6.49)$$

at equilibrium. Also at equilibrium

$$\phi_{\alpha\beta}\binom{ll'}{ss'} = \left[\frac{\partial^2 V}{\partial u_\alpha\binom{l}{s} \partial u_\beta\binom{l'}{s'}}\right]_0. \quad (6.50)$$

In the harmonic approximation, we neglect all higher-order terms. V_0 is a static term and can be ignored. In addition, $V_1 = 0$, because $\phi_\alpha\binom{l}{s}$ is the negative of the force acting on the atom ls and this is zero at equilibrium. Therefore, the total Hamiltonian for harmonic motion is

$$V = V_0 + \frac{1}{2}\sum_{ls\alpha} M_s u_\alpha^2\binom{l}{s} + \frac{1}{2}\sum_{\substack{ls \\ l's'}}\sum_{\alpha\beta} \phi_{\alpha\beta}\binom{ll'}{ss'} u_\alpha\binom{l}{s} u_\beta\binom{l'}{s'}. \tag{6.51}$$

The equation of motion for atom ls then is

$$M_s \ddot{u}\binom{l}{s} = -\frac{\partial V}{\partial u\binom{l}{s}} = -\sum_{l's'} \phi\binom{ll'}{ss'} u\binom{l'}{s'}. \tag{6.52}$$

This gives an infinite number of coupled differential equations to solve. This can be simplified by making use of lattice periodicity. A suitable wave-like solution is then (note the similarity to the way we treated the linear chains)

$$u\binom{l}{s} = M_s^{-\frac{1}{2}} U_s e^{i\omega t} e^{i\mathbf{k}\cdot\mathbf{r}(ls)}. \tag{6.53}$$

where $\mathbf{r}\,(ls)$ is the position vector of the sth atom in the lth unit cell. Therefore,

$$\begin{aligned}M_s \ddot{u}\binom{l}{s} &= -\omega^2 M_s^{\frac{1}{2}} U_s e^{i\omega t} e^{i\mathbf{k}\cdot\mathbf{r}(ls)} \\ &= \sum_{l's'} \phi\binom{ll'}{ss'} M_{s'}^{-\frac{1}{2}} U_{s'} e^{i\omega t} e^{i\mathbf{k}\cdot\mathbf{r}(l's')}.\end{aligned} \tag{6.54}$$

Rearranging and reintroducing directions α and β

$$\begin{aligned}\omega^2 U_{\alpha s} &= \sum_{\beta l's'} \phi\binom{ll'}{ss'} \frac{1}{\sqrt{M_s M_{s'}}} U_{\beta s'} e^{i\omega t} e^{i\mathbf{k}\cdot[\mathbf{r}(l's')-\mathbf{r}(ls)]} \\ &= \sum_{\beta s'} D_{\alpha\beta}(ss',\mathbf{k}) U_{\beta s'}.\end{aligned} \tag{6.55}$$

$D(ss', \mathbf{k})$ is known as the *dynamical matrix*. In this way, the problem is reduced to the solution of $3n$ equations, where n is the number of atoms in a *primitive* unit cell. A non-trivial solution is found when

$$|D_{\alpha\beta}(ss', \mathbf{k}) - \omega^2 \delta_{\alpha\beta} \delta_{ss'}| = 0. \tag{6.56}$$

The dynamical matrix is Hermitian, and so the eigenvalues are real. The result is that we obtain $3n$ solutions for ω^2 for any value of $\mathbf{k}$. By diagonalizing the dynamical matrix, the problem is then re-specified by new coordinates, the so-called *normal coordinates*. The resulting solutions then correspond to $3n$ independent *normal modes*.

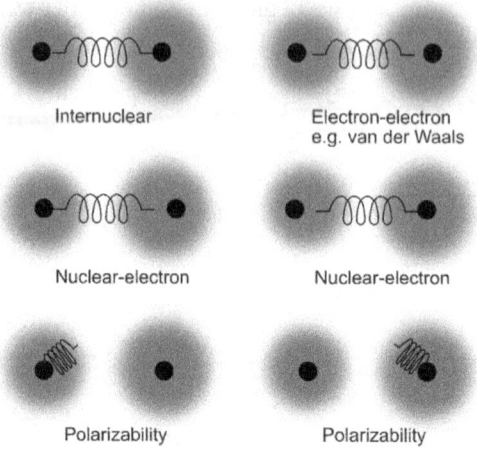

Figure 6.20. The force constant information needed to describe a diatomic molecule in full.

In principle, solving the full lattice dynamics problem should be straightforward for any crystal structure, and indeed computer programs are now available that attempt to do just that. However, in reality the task is even more complex, because the force constants are of variable origin and are often unknown. Just consider, for example, the trivial case of a diatomic molecule consisting of two different atoms (figure 6.20). It can be seen that, even in this simple molecule, we require a great deal of information to fully determine the frequencies (the eigenvalues) of vibration for the individual atoms, although by group-theoretical arguments we can easily use symmetry to determine the directions of movement (the eigenvectors). In this example, we know that the mode of vibration is simply described by the two atoms moving towards and away from each other. Determination of the frequency, however, requires knowledge of at least the most important force constants, so it is not trivial after all!

Figure 6.21 shows a couple of examples where the dispersion curves have been calculated using an *ab initio* force constant model and compared with measurements for two relatively easy cases, diamond (all-face-centred cubic) and graphite (hexagonal). The curves are shown for different paths in the Brillouin zones and the letters indicate the critical points. The agreement between measured and calculated curves is excellent in these cases, but with crystal structures of lower symmetry it may be more difficult to obtain such good agreement.

6.6 Heat conduction

Another thermal process in which reciprocal space plays a significant role is thermal conduction. It is obvious that phonons carry thermal energy across the solid, but what is it that results in some materials conducting heat less quickly than others? Why are some materials, like diamond, good thermal conductors, while others are good thermal insulators? To explain this, we need to understand that thermal energy is conducted through the material by two types of quantum particle, phonons and

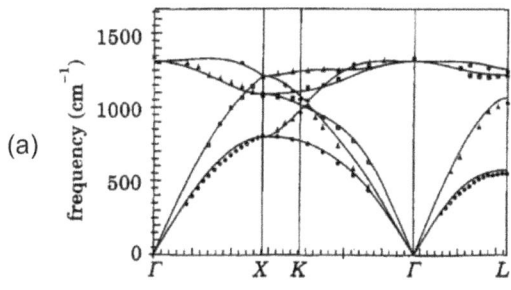

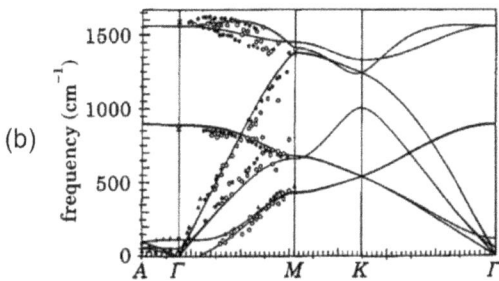

Figure 6.21. Observed and calculated phonon dispersion curves for (a) diamond and (b) graphite. Reproduced with permission from [7].

electrons. If we restrict ourselves to electrically insulating materials, then it is principally phonons that are responsible for the heat conduction. In order now to limit phonon heat conduction, in other words create thermal resistivity, we need a mechanism whereby the flow of phonons is in some way impeded. One possibility is that impurities and imperfections within the crystal structure will tend to impede phonon propagation. But the other important mechanism, which occurs even in the purest of crystalline materials, is through phonon–phonon collisions. There are two collision-dominated regimes that need to be considered.

Normal processes

Consider two phonons with wave-vectors $\mathbf{k}_1$ and $\mathbf{k}_2$ travelling from left to right with respect to our choice of axes.

Figure 6.22(a) shows what happens when the two phonons interact to create a third phonon:

$$\mathbf{k}_1 + \mathbf{k}_2 = \mathbf{k}_3. \tag{6.57}$$

or in terms of momentum conservation

$$\hbar\mathbf{k}_1 + \hbar\mathbf{k}_2 = \hbar\mathbf{k}_3. \tag{6.58}$$

such that wave-vector $\mathbf{k}_3$ still lies within the 1st Brillouin zone. In other words, a collision between two real waves with horizontal components travelling from left to

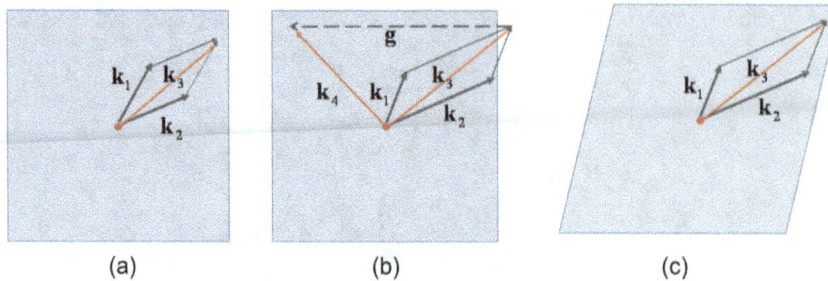

Figure 6.22. (a) normal process (b) Umklapp (c) Umklapp or not? A different choice of Brillouin zone.

right creates another real wave also with a horizontal component travelling from left to right. Phonon collisions are examples of anharmonic interactions which we have so far ignored. The effect of the collision is to contribute a resistance to the flow of thermal energy. This type of collision is called a *normal process*.

Umklapp processes

There is a more important, although less well understood, type of phonon collision that needs to be considered as a mechanism limiting heat propagation. Once again (figure 6.22(b)) consider two phonons with wave-vectors k_1 and k_2 having horizontal components travelling from left to right. As before, they interact to produce a third phonon with wave-vector given by

$$k_3 = k_1 + k_2. \tag{6.59}$$

However, this time the initial wave-vectors are sufficiently large that the new phonon wave-vector lies outside the 1st Brillouin zone and is therefore not a physically real solution for a wave with a component travelling to the right. We can now appeal to the translational periodicity of the reciprocal lattice and via the Bloch Theorem subtract a reciprocal lattice vector g from k_3. The result is now a wave-vector k_4 that lies within the 1st Brillouin zone but has its horizontal component now pointing from right to left:

$$k_4 = k_1 + k_2 - g. \tag{6.60}$$

So, two waves travelling from left to right collide and produce a new wave that has effectively been reflected back. Such a collision therefore has the effect of reversing the horizontal flow of thermal energy and thus in this case a large thermal resistivity is obtained. This is known as an *Umklapp process*.

We need, however, to inject a word of caution here. Recall that a Brillouin zone is simply a unit cell in reciprocal space [8] and so it can be defined in an infinite number of ways. The 1st Brillouin zone in figures 6.22(a) and (b) has been drawn using the Wigner–Seitz construction, perfectly acceptable given the high symmetry of the reciprocal lattice in the figures. But suppose we choose a different shape for our unit cell (figure 6.22(c)). This time the wave-vector k_3 lies within this definition of the Brillouin zone (see Cracknell [9]). So what has happened to Umklapp? As pointed

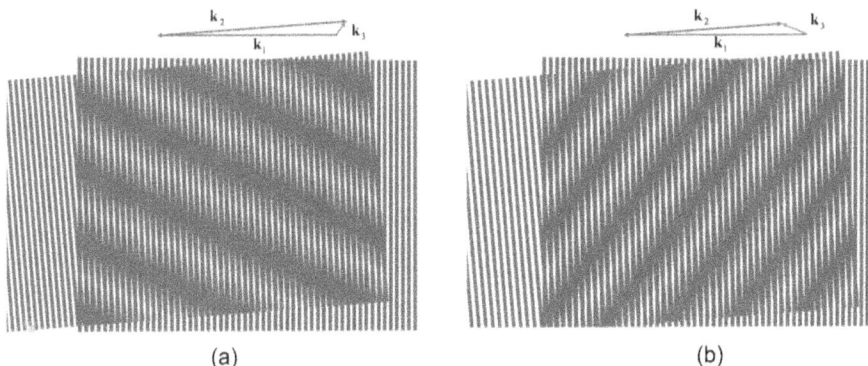

Figure 6.23. Moiré fringe technique to illustrate (a) a normal process (b) an Umklapp process.

out by Cracknell, in the past those writers who are careful about this matter have appreciated that a change in the choice of the Brillouin zone will alter the division into normal and Umklapp processes.

We have treated the two processes, normal and Umklapp, in terms of the interactions between quantized quasiparticles. However, it is educational to look at this in terms of waves interfering with one another. A nice to way to illustrate is to use a Moiré fringe technique. Figure 6.23(a) shows how a normal process can be 'simulated'. Two sets of fringe patterns of different wavelength and orientation are superimposed and a series of dark fringes can be seen where the two sets of fringes interfere. The diagram also shows the triangle of wave-vectors given by (6.59). In this case k_3 has a horizontal component from left to right. Figure 6.23(b) shows a simulation of the Umklapp process: in this case the result is a wave with a horizontal wave-vector component from right to left. What this way of visualizing these processes shows is that the interaction between the phonons can be treated simply as the interference of waves, in which case it can be seen that it is possible to interfere two waves travelling from left to right and yet have a result travelling in the opposite direction. It reminds me of the old cowboy movies (remember those?) in which the spokes of the wagon wheels seem to travel the opposite direction to the stagecoach, as a result of interference with the camera shutter cycle.

Now that we understand the difference between normal and Umklapp processes, let us look at a particular case. The mineral diamond is unusual in that, unlike a metal, it is an excellent electrical insulator but a superb thermal conductor. In fact, it is such a good thermal conductor that it is sometimes used as a heat sink for electronic components. By the way, a way to distinguish between a genuine and a fake diamond is to place it on the tongue. A genuine diamond should feel cold like a metal. The question we need to address here is: why is diamond such a good thermal conductor but at the same time a good electrical insulator? The following statements answer this:

- Diamond has covalent bonding between carbon atoms.
- This makes diamond very hard, as bonds are very stiff.
- Therefore, diamond has a high Debye temperature.

- At room temperature diamond is far below the Debye temperature.
- This means that most phonons at room temperature are of low frequency, long wavelength.
- So, most of the phonons in diamond at room temperature have short wave-vectors.
- Therefore, there are few opportunities for Umklapp to occur.
- The covalent bonding means that the electrons are tightly bound, so that diamond is a good electrical insulator.
- Note that metals are good electrical and thermal conductors, because then heat is transported by both electrons and phonons.

6.7 Interaction with radiation

In this section, we shall consider some of the ways in which phonon dispersion information can be obtained. Consider figure 6.24.

The three principal methods to discuss are optical (infrared) absorption, Raman scattering and neutron inelastic scattering. Typical optical wavelengths are in the region of 3000–8000 μm and so typical wave-vectors are between about 1.2×10^{-4} and 3×10^{-4} Å^{-1}, to be compared with a typical reciprocal lattice distance of about 1 Å^{-1}. Therefore, optical wavelengths can be used to probe phonons with $k \approx 0$, where the phonon wavelengths are very large. This can be done by optical absorption spectroscopy or by inelastic light scattering. In the latter case, there are two regimes to consider: Raman scattering will enable the optic modes close to $k = 0$ to be studied, whilst Brillouin scattering will cover the acoustic modes. The main way of studying phonons at higher wave-vectors is neutron inelastic scattering. This works because thermal neutrons have energies in the meV range, and this is comparable with typical phonon energies. In addition, typical thermal neutron wavelengths are in the Ångström range and so most of the Brillouin zone can be accessed.

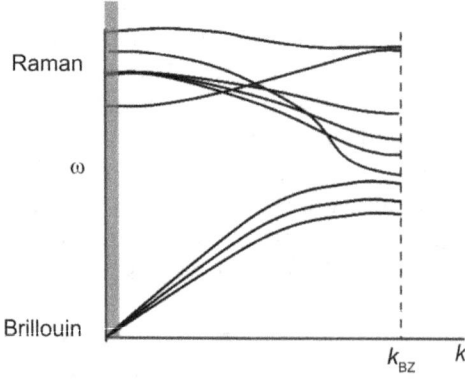

Figure 6.24. Regions of the phonon dispersion curves accessible to different radiations (reduced zone scheme).

Absorption spectroscopy

Typically, infrared radiation is used, since optic phonon frequencies tend to lie in this region of the electromagnetic spectrum. A transition between an initial quantum state ϕ_i and a final state ϕ_f is only possible when the quantum matrix element

$$\langle \phi_f | \hat{H} | \phi_i \rangle \neq 0. \tag{6.61}$$

In optical absorption spectroscopy, the electric field $\mathbf{E}(t)$ of the light couples with an oscillating dipole $\mathbf{p}(t)$ resulting in a time-dependent Hamiltonian. Therefore, the transition occurs for

$$\langle \phi_f | \hat{H}' | \phi_i \rangle \neq 0. \tag{6.62}$$

where $\hat{H}'$ represents the additional energy acquired when the light is absorbed. Therefore,

$$\langle \phi_f | \hat{H}' | \phi_i \rangle = \langle \phi_f | \mathbf{p}(t) \cdot \mathbf{E}(t) | \phi_i \rangle = e \langle \phi_f | \mathbf{r}(t) \cdot \mathbf{E}(t) | \phi_i \rangle \neq 0. \tag{6.63}$$

The vector $\mathbf{r}(t)$ is the time-dependent position vector for an atom. Provided that the atom is charged, i.e., an ion, then this means that an oscillating dipole exists and absorption can take place.

This picture is a simplified description of the selection rule that allows infrared absorption to occur. For instance, alkali halides are ionic crystals and in the optic modes the positive cations move towards and away from the negative anions thus creating oscillating dipoles. The consequence is that there is a large absorption of light at a frequency matching the relevant optic mode frequency. However, note that diamond is not ionic and yet a certain amount of weak absorption is seen in the infrared. This has been much discussed in the literature, with the suggestion that it arises either from impurities or defects or from two-photon absorption. The latter manifests itself as a secondary effect caused by the deformation of the charge distribution during vibration leading to a second-order dipole moment.

Inelastic scattering of light

The two optical inelastic scattering techniques to be discussed are Raman and Brillouin scattering [10]. In both cases a monochromatic optical beam, typically from a laser, is incident on the sample and the frequency spectrum is analyzed at some angle from the incident beam direction. To understand the process involved, first consider a light wave with an oscillating electric field $\mathbf{E}$ given by

$$\mathbf{E} = \mathbf{E}_0 e^{i(\mathbf{k}_0 \cdot \mathbf{r} - \omega t)}. \tag{6.64}$$

On encountering an atom in the crystal this electric field polarizes the electron cloud thus *inducing* a dipole moment (figure 6.25). The polarization at any point in time then will be given by

$$\mathbf{P} = \alpha \mathbf{E}. \tag{6.65}$$

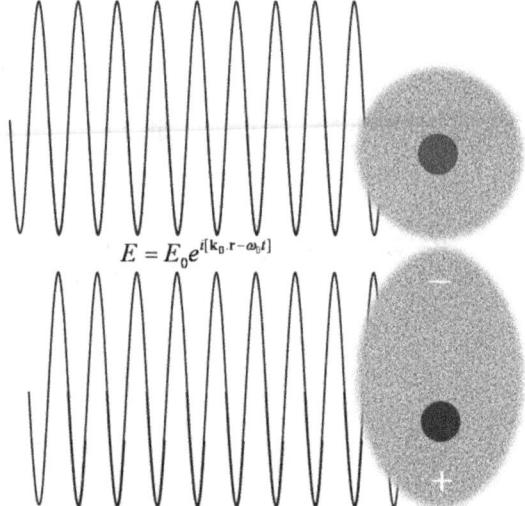

Figure 6.25. The effect of an incident light wave on an atom is to momentarily displace the electron cloud around the nucleus thus inducing an electric dipole.

where α is the electronic polarizability. This is actually given by a 2nd-rank tensor:

$$P_l = \alpha_{lm} E_m. \tag{6.66}$$

so that α_{lm} creates a polarization in direction l when the optical field is polarized along direction m. In this case the matrix selection rule is determined by

$$\langle \phi_f | \hat{H}' | \phi_i \rangle = \langle \phi_f | \hat{\alpha} | \phi_i \rangle. \tag{6.67}$$

where $\hat{\alpha}$ is the polarizability operator.

Now the atom in the crystal is not static, but is oscillating with a normal coordinate given by

$$\mathbf{U}_j = \mathbf{U}_j^0 e^{i(\mathbf{k}_j \cdot \mathbf{r} - \omega_j t)}. \tag{6.68}$$

for the jth phonon. The effect of this is to make the polarizability itself dependent on this normal coordinate:

$$\alpha = \alpha(\mathbf{U}_j). \tag{6.69}$$

Expanding in a Taylor series about the static polarizability $\alpha(0)$

$$\alpha = \alpha(0) + \sum_j \left(\frac{\partial \alpha}{\partial \mathbf{U}_j}\right)_0 \mathbf{U}_j + \frac{1}{2} \sum_j \sum_{j'} \left(\frac{\partial^2 \alpha}{\partial \mathbf{U}_j \partial \mathbf{U}_{j'}}\right)_0 \mathbf{U}_j \mathbf{U}_{j'} + \ldots \tag{6.70}$$

Multiplying by the electric field and substituting by (6.68) we get

$$\mathbf{P} = \alpha(0)\mathbf{E}_0 e^{i(\mathbf{k}_0 \cdot \mathbf{r} - \omega_0 t)}$$
$$+ \sum_j \left(\frac{\partial \alpha}{\partial \mathbf{U}_j}\right)_0 \mathbf{E}_0 \mathbf{U}_j^0 e^{i[(\mathbf{k}_0 \pm \mathbf{k}_j) \cdot \mathbf{r} - (\omega_0 \pm \omega_j)t]} \quad (6.71)$$
$$+ \frac{1}{2}\sum_j \sum_{j'} \left(\frac{\partial^2 \alpha}{\partial \mathbf{U}_j \partial \mathbf{U}_{j'}}\right)_0 \mathbf{E}_0 \mathbf{U}_j^0 \mathbf{U}_{j'}^0 e^{i[(\mathbf{k}_0 \pm \mathbf{k}_j \pm \mathbf{k}_{j'}) \cdot \mathbf{r} - (\omega_0 \pm \omega_j \pm \omega_{j'})t]} + \ldots$$

The scattered intensity is then given by

$$I \propto PP^*. \quad (6.72)$$

Let us look at each of the terms in (6.71) in turn. The first term represents the light scattered with the same frequency as the incident light frequency, and therefore corresponds to straightforward elastic scattering. In the second term, we see that the incident light momentum has been changed according to

$$\hbar \mathbf{k}_s = \hbar \mathbf{k}_0 \pm \hbar \mathbf{k}_j. \quad (6.73)$$

and the light energy according to

$$\hbar \omega_s = \hbar \omega_0 \pm \hbar \omega_j. \quad (6.74)$$

We see from this that the incident light frequency has increased by destroying a phonon j in the process or decreased by creating a phonon j in the process (the two regimes are known as the antiStokes and Stokes regimes, respectively). In Raman scattering it is the optic modes that are found symmetrically either side of the incident beam frequency (figure 6.26).

The Raman spectrum therefore consists of a central very strong elastic peak (actually it does have some width and so is quasielastic), and on either side, peaks

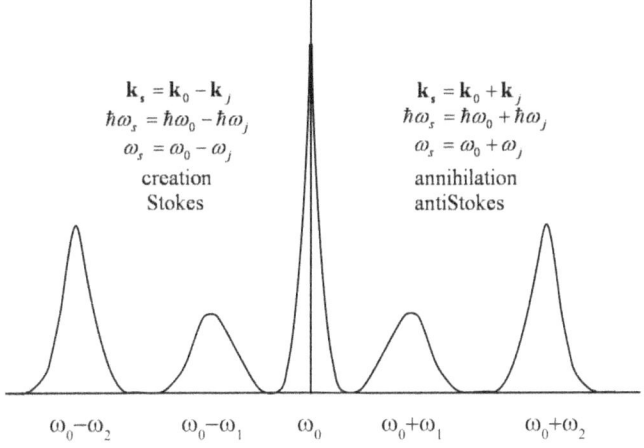

Figure 6.26. Schematic diagram of Raman scattering.

corresponding to the optic mode frequencies close to $k_j \to 0$. If now the frequency scale is stretched out, it is seen that the central peak becomes broader and in practice shows some extra features. These are due to the low-frequency acoustic modes and this type of scattering is usually known as Brillouin scattering. This is useful for studying the elastic properties of a crystal, since the acoustic modes are connected with unit cell distortions, which therefore affect the whole crystal. Note that one can make use of polarizers placed in the incident and scattered beam paths to cut out or allow through different phonon peaks, thus helping one to identify the types of phonons found in the spectrum. The third term in (6.71), which is generally weaker than the previous two, involves changes in wave-vector and light-frequency of the form

$$\mathbf{k}_s = \mathbf{k}_0 \pm \mathbf{k}_j \pm \mathbf{k}_{j'}$$
$$\omega_s = \omega_0 \pm \omega_j \pm \omega_{j'}. \tag{6.75}$$

This is therefore a two-phonon interaction with the light. Because terms such as $\mathbf{k}_j \pm \mathbf{k}_{j'}$ can now be comparable with $\mathbf{k}_0$, the wave-vector for the incident light, the response signal covers phonon information from the whole of the Brillouin zone. It is usually seen as a background to the first order scattering and provides some information on the phonon density of states rather than on individual phonons.

Inelastic scattering of neutrons

The most general method for obtaining phonon dispersion diagrams is by using the inelastic scattering of thermal neutrons [11]. As with the optical method, a monochromatic beam is often used and a measure of the change in energy is obtained. To do this use is made of a triple-axis diffractometer (figure 6.27). In this system [12], a beam of thermal neutrons from, say, a nuclear reactor is first monochromatized by a (111) cut crystal of germanium. This beam is incident on the crystal to be studied. The scattered neutrons are then energy-analyzed by a second crystal and a detector. By combination of rotations of the sample, analyzer crystal and the detector, different regions of the Brillouin zone can be reached. Figure 6.28 shows an example in which an incident beam with wave-vector $\mathbf{k}_i$ is incident at some direction with respect to the reciprocal lattice, and the wave-vector $\mathbf{k}_f$ is of the correct length and orientation to measure data between Γ and X points of a cubic Brillouin zone. In this example, note that we do not have to measure at the X-point at the closest position to the origin of reciprocal space (e.g. at (0, ½, 0)), but, instead we can choose the X-point in a neighbouring Brillouin zone, in this case at (1, ½, 0). Indeed, it may not be possible to measure at the nearest X-point, as this depends on the initial neutron energy and the phonon energy. In this example, the final wave-vector is greater than the initial wave-vector:

$$\mathbf{k}_f - \mathbf{k}_i = \mathbf{Q} = \mathbf{q} + \mathbf{g}. \tag{6.76}$$

$$Q^2 = k_1^2 + k_2^2 - 2k_1 k_2 \cos 2\theta. \tag{6.77}$$

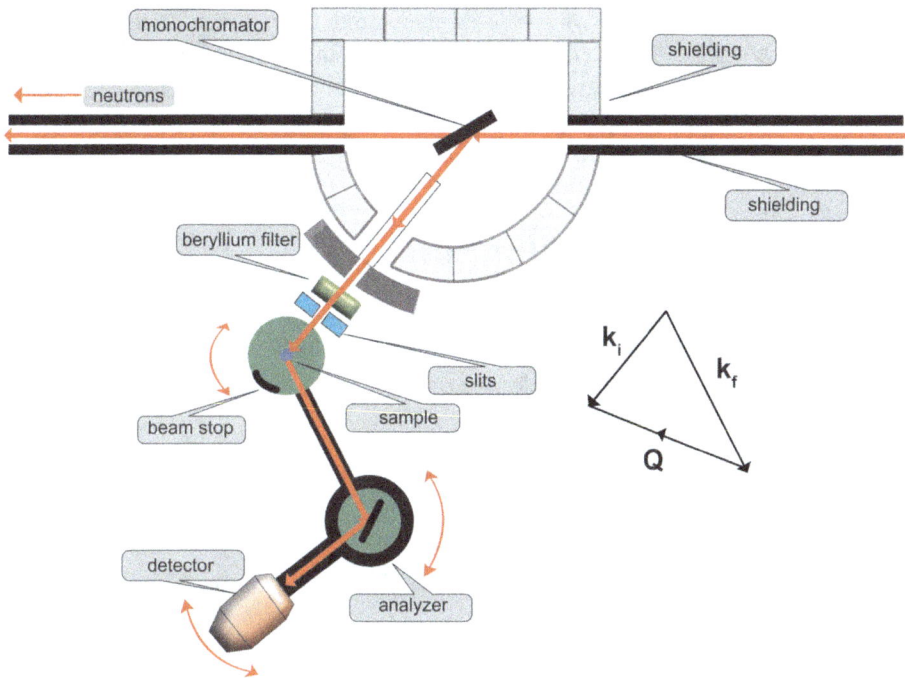

Figure 6.27. The triple-axis spectrometer.

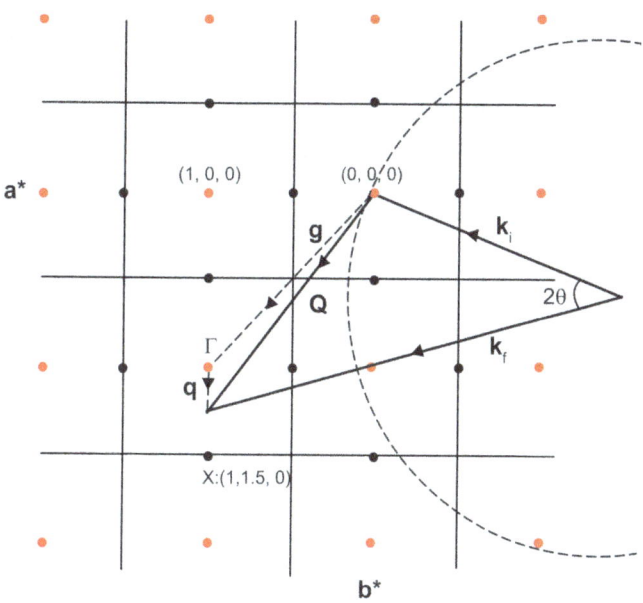

Figure 6.28. Measurement of phonons along Γ–X-direction of the Brillouin zone. Red: reciprocal lattice. Black dot: X-points.

The vector **q** is the phonon wave-vector measured from the Γ point at (1, 1, 0) towards the X-point. Here, the initial neutron energy has been increased by destroying a phonon. The initial and final wave-vectors are given by

$$k_i = \sqrt{E_i \frac{2m}{\hbar^2}}$$
$$k_f = \sqrt{(E_i + E_{\text{phonon}})\frac{2m}{\hbar^2}}. \quad (6.78)$$

where m is the neutron mass. With a different orientation of the crystal it is possible for the neutron to lose energy, thus creating a phonon. In this way, it is possible to trace out the phonon dispersion along any line in the Brillouin zone (as seen for example in figure 6.21).

6.8 Free electrons in a metal

One shouldn't work on semiconductors, that is a filthy mess; who knows whether any semiconductors exist.

Wolfgang Pauli

I now end by giving a brief introduction to the electronic properties of metals, insulators and semiconductors and how they can be explained in terms of reciprocal space concepts. More details can be found in the many standard solid-state physics textbooks available.

The conductivity of electricity in metals is governed by the flow of electrons. To a good approximation, the electrons that are important are the valence electrons supplied by the metal atoms which are free to move without interactions with the inner electrons around the nuclei. The energy of these free electrons is given by

$$E = \frac{\hbar^2 k^2}{2m}. \quad (6.79)$$

So, the energy curve follows a parabola (figure 6.29). Since electrons are by nature fermions, they do not occupy the same energy states at the same time. This means

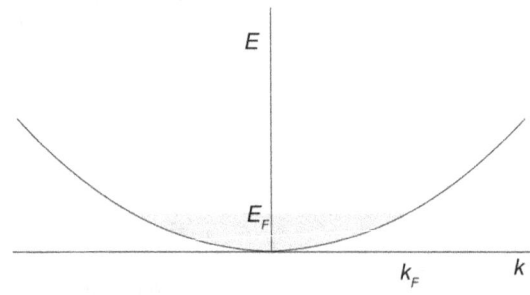

Figure 6.29. Energy versus wave-vector.

that the conducting electrons occupy levels from a baseline (set at 0 here) up to a maximum value of energy E_F, the *Fermi energy*.

Note *The Fermi energy is the energy of the maximum occupied state at **absolute zero of temperature**.*

The Fermi wave-vector is given by

$$k_F = \left(\frac{3\pi^2 N_{\text{states}}}{V_{\text{crystal}}}\right)^{1/3}. \tag{6.80}$$

In many books, N_{states}, the number of occupied states, is often replaced by Ne, the total number of electrons in the metal. However, as we shall see later, this can result in a misunderstanding of the Brillouin zone. The Fermi energy is

$$E_F = \frac{\hbar^2}{2m}\left(\frac{3\pi^2 N_{\text{states}}}{V_{\text{crystal}}}\right)^{2/3}. \tag{6.81}$$

The Fermi energy describes a shape in **k**-space, the surface of which is known as the *Fermi surface*. For a monovalent metal, this is in the form of a sphere, while more complex shapes can be found in other metals. Now, the Fermi–Dirac function is given by

$$f = \frac{1}{e^{(E-\mu)/k_B T} + 1}. \tag{6.82}$$

where μ is the chemical potential. At a temperature of 0 K this is equal to the Fermi energy. Now, it turns out that the Fermi energy for a typical metal is in the few eV range (e.g. 7 eV for metallic copper), amounting to Fermi temperatures of tens of thousands of degrees Kelvin. At room temperature, therefore, we are well below the Fermi temperature, and so we can effectively replace the chemical potential by the Fermi energy:

$$f \simeq \frac{1}{e^{(E-E_F)/k_B T} + 1}. \tag{6.83}$$

The density of states for a three-dimensional metal is given by

$$D(E) = \frac{V_{\text{crystal}}}{2\pi^2}\left(\frac{2m}{\hbar}\right)^{3/2} E^{1/2}. \tag{6.84}$$

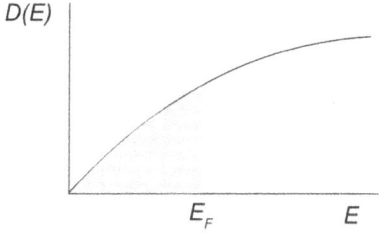

Figure 6.30. Density of states versus energy for free electrons.

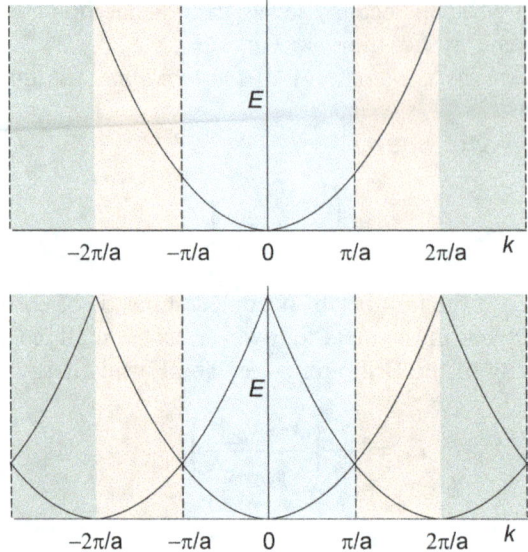

Figure 6.31. Effect of translational periodicity on 'almost' free electron energy plots.

where a factor of two has been included to deal with the fact that electrons occur in two spin states. We see that $D(E)$ is parabolic in E (figure 6.30).

We now introduce translational symmetry into our discussion. Figure 6.31 (top) now introduces Wigner–Seitz unit cells to form conventional extended Brillouin zones. In the bottom diagram, the vertical dashed lines mark the repeating unit cells and within each unit cell the E–k curves are folded in to create a repeating pattern.

From this, the reduced zone $0 \leqslant k \leqslant \pi/a$ can be used instead, showing an energy level diagram at any value of k.

6.9 Nearly free electrons

In a real crystalline solid, the conduction electrons responsible for current flow experience forces exerted on them by their surroundings, mainly the atomic nuclei and inner electrons. They therefore are subject to some sort of periodic potential (figure 6.32), analogous to the way that thermal waves are influenced by the periodic array of force constants in the crystal. For simplicity, I shall consider this in one dimension in order to illustrate the way in which this problem can be handled.

Consider the wave-function at the Brillouin zone boundary. From figure 6.31 (bottom) we see that the solutions are two-fold degenerate in energy. This arises from the fact that at the zone boundary the electron wave can be thought of as being Bragg-reflected into the opposite direction. We can therefore write for the two solutions

$$\begin{aligned} \psi(+) &= Ae^{i\pi x/a} + Ae^{-i\pi x/a} = 2A\cos \pi x/a \\ \psi(-) &= Ae^{i\pi x/a} - Ae^{-i\pi x/a} = 2Ai\sin \pi x/a. \end{aligned} \tag{6.85}$$

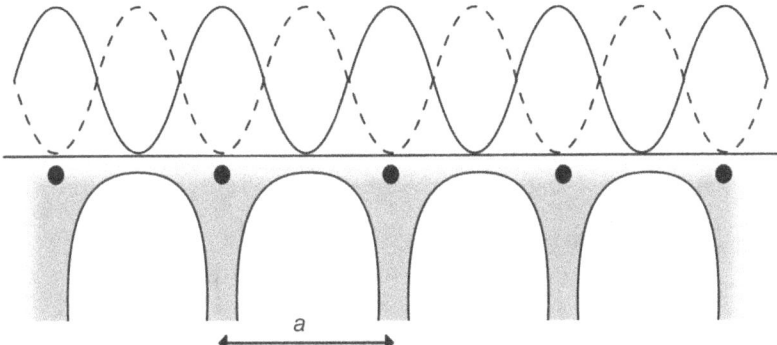

Figure 6.32. Periodic lattice potential (bottom) and probability density waves (top). Full line $\Psi(+)^2$, dashed line $\Psi(-)^2$.

The probability densities then are given by

$$\psi(+)\psi(+)^* = 4\cos^2 \pi x/a$$
$$\psi(-)\psi(-)^* = 4\sin^2 \pi x/a. \tag{6.86}$$

We see that the + probability wave peaks on the nuclei, while the − probability wave peaks half-way between. This means an electron occupying one solution is subject to a different part of the periodic potential than the other, thus lifting the two-fold degeneracy, and hence creating an energy gap.

More generally, this periodic potential can be expressed in one dimension via a Fourier series thus

$$V(x) = \sum_g V_g e^{igx}. \tag{6.87}$$

We now use the following Bloch wave for any general value of k

$$\psi = A e^{ikx} + B e^{i(k-g)x}. \tag{6.88}$$

Subtracting g in the exponent for the second term reverses the sense of the wave, so that this wave-function represents two opposing running waves. Substituting into the Schrödinger equation

$$-\frac{\hbar^2}{2m}\frac{\partial^2 \psi}{\partial x^2} + V\psi = E\psi$$

$$\therefore -\frac{\hbar^2}{2m}[-Ak^2 e^{ikx} - B(k-g)^2 e^{i(k-g)x}]$$

$$+ \left(\sum_g V_g e^{igx}\right)(A e^{ikx} + B e^{i(k-g)x}) \tag{6.89}$$

$$= E(A e^{ikx} + B e^{i(k-g)x}).$$

We now compare coefficients in e^{ikx} and $e^{i(k-g)x}$ to get two equations

$$e^{ikx} \Rightarrow A\frac{\hbar^2}{2m}k^2 + AV_0 + BV_g = EA$$
$$e^{i(k-g)x} \Rightarrow B\frac{\hbar^2}{2m}(k-g)^2 + BV_0 + AV_{-g} = EB.$$
(6.90)

For a non-trivial solution

$$\begin{vmatrix} \frac{\hbar^2}{2m}k^2 + V_0 - E & V_g \\ V_{-g} & \frac{\hbar^2}{2m}(k-g)^2 + V_0 - E \end{vmatrix} = 0.$$
(6.91)

This gives the following result

$$\left[\frac{\hbar^2}{2m}k^2 + V_0 - E\right]\left[\frac{\hbar^2}{2m}(k-g)^2 + V_0 - E\right] - V_g V_{-g} = 0.$$
(6.92)

At the zone boundary we can substitute $k_{ZB} = g/2$ and so, after taking the square root

$$E = V_0 + \frac{\hbar^2 k_{ZB}^2}{2m} \mp |V_g|.$$
(6.93)

The term V_0 can be ignored for our purposes, as it simply acts as a base value for the energy. The second term is just the solution for the free electron case at the zone boundary. However, the third term represents a perturbation to the free electron solution by causing the two-fold degeneracy to split into two separate energies separated by $2V_g$ (figure 6.33).

From this we see that the E–k plot is split into two bands: the upper one is called the *conduction band*, the lower is called the *valence band*. Between the two, there is a forbidden region for the electrons known as the *band gap*. It is the presence of this band gap that determines the electrical conduction properties of the solid. Figure 6.34 shows how the extended zone scheme translates into the reduced zone scheme

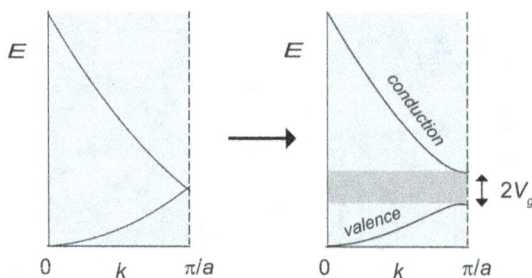

Figure 6.33. Reduced zone. Creation of an energy band gap (grey) by a first-order perturbation by the lattice potential.

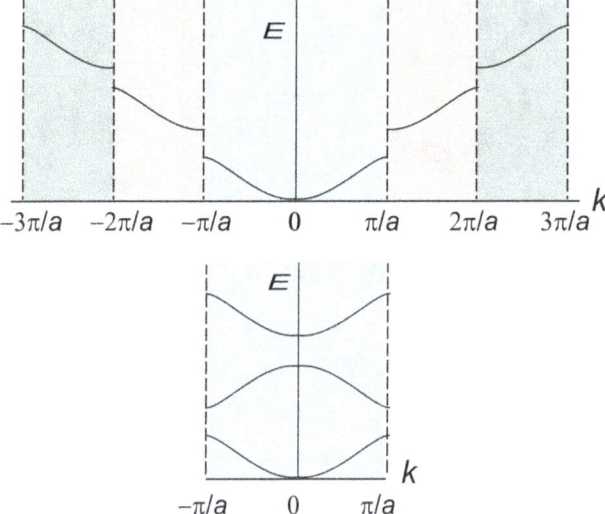

Figure 6.34. The extended zone scheme (top) and its associated reduced zone scheme.

where the different bands are more obvious as a series of energy levels for different values of the wave-vector.

So, if the band gap is large, electrons are not easily promoted from the valence to the conduction band, in which case the solid acts an electrical insulator. If the band gap is smaller, then this becomes an intrinsic semiconductor, as then by Boltzmann statistics thermal energy helps to promote the electrons across the gap. An intrinsic semiconductor therefore increases its electrical conductivity as a function of temperature. If the band gap is negligible or even negative then metallic behaviour is expected.

Band structure diagrams for real solids are more complicated than the simple picture given so far. The details of the bands and how the energy depends on wave-vector depend on the complexity of the actual lattice potential terms present in any particular case. These days, realistic model potential functions can now be computed and the fit between experimental and theoretical band structures can be excellent. The complexity in real band structures also arises from the overlap between the various types of atomic orbitals of the atoms. The band structure for copper metal (figure 6.35) illustrates this, where we see various bands formed by overlap of 3s, 3p and 3d orbitals. The Fermi surface for copper is seen to be spherical but with extra projections towards the L-point of the Brillouin zone boundary.

6.10 Metal or insulator?

Relatively simple concepts can be used to determine if a particular solid should show metallic or insulator behaviour, by counting the number of electrons available within the different electronic bands. You will see in many textbooks that it is said that metallic behaviour occurs when an atom in an elemental solid has an odd number of electrons per atom. However, this statement needs a proper examination,

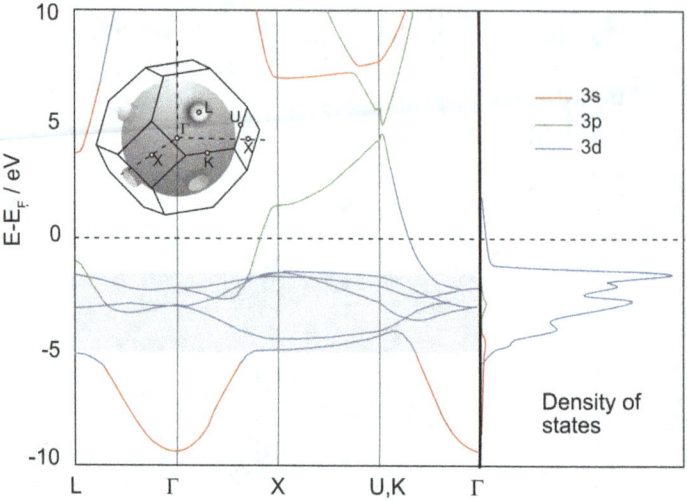

Figure 6.35. Band structure, density of states and Fermi surface for metallic copper. The grey band indicates s–d orbital hybridization.

as we need to be careful to compare unit cells in direct space with unit cells in reciprocal space in order to be able to count correctly the number of occupied states in an energy band.

Note that our use of the Wigner–Seitz construction to define the Brillouin zone as a primitive reciprocal space unit cell means that we need to consider also primitive unit cells in direct space. The number of independent states in the Brillouin zone is given by

$$N_{\text{states}} = \frac{V_{BZ}}{V_{\text{state}}}. \tag{6.94}$$

where V_{BZ} is the volume of the Brillouin zone. For a cubic crystal with lattice spacing a

$$V_{BZ} = \left(\frac{2\pi}{a}\right)^3 = \frac{8\pi^3}{V_{\text{cell}}}. \tag{6.95}$$

Here, V_{cell} is the volume of a primitive unit cell in real space. V_{state} is the volume around any **k**-state.

$$\therefore V_{\text{state}} = \left(\frac{2\pi}{L}\right)^3 = \frac{8\pi^3}{V_{\text{crystal}}}$$
$$\therefore N_{\text{states}} = \frac{8\pi^3}{V_{\text{cell}}} / \frac{8\pi^3}{V_{\text{crystal}}} = \frac{V_{\text{crystal}}}{V_{\text{cell}}}. \tag{6.96}$$

and so the number of states equals the number of primitive unit cells in the crystal. Since we are dealing with electrons the number of available states per band is equal to $2N_{states}$, taking account of the spin degeneracy of an electron. To illustrate this, consider some simple cases.

Sodium

Sodium is a metal that crystallizes in a body-centred cubic structure with two Na atoms per bcc unit cell. As the conventional choice of a Wigner–Seitz construction for the Brillouin zone gives us a primitive unit cell in reciprocal space, we need to consider the number of Na atoms in the direct space *primitive* unit cell too.

We know that in the Brillouin zone there are $2N_{states}$ available states per band. In the primitive direct space unit cell there is one Na atom (the volume of the bcc cell is twice that of the primitive cell). Now each Na atom has electronic configuration $1s^2 2s^2 2p^6 3s^1$, and so for conductivity we consider the outer $3s^1$ electrons. If there are $2N_{states}$ available, then only N_{states} are occupied at a temperature of absolute zero. This means that the $3s$ band is half-occupied and so the Fermi energy lies well below the top of the band where a band gap can occur (see figure 6.36 where the Fermi surface is close to spherical and lies well inside the Brillouin zone). Note that by the Pauli exclusion principle, only those electrons close to the Fermi energy will be available to conduct electricity, thus making sodium a metal.

But suppose we invented a hypothetical sodium crystal structure with, say, four Na atoms per bcc unit cell. We would then have two Na atoms per primitive cell, contributing two electrons per unit cell. We would then occupy all $2N_{states}$, and therefore have a Fermi energy just below the band gap. If this gap were large, then we would conclude that sodium is an insulator! So, just because the atoms have an odd number of electrons, it does not follow theoretically that it must be a metal. What matters is whether there is an odd number of electrons per primitive unit cell, rather than per atom!

Calcium

Calcium is also a metal, even though it has an even number of electrons. It crystallizes with four Ca atoms in an fcc unit cell. Therefore, there is one Ca atom per primitive unit cell. Calcium is divalent and so each primitive unit cell contributes

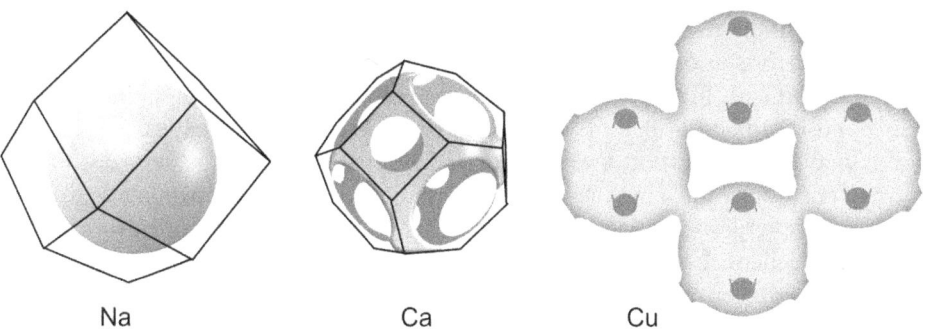

Figure 6.36. Fermi surfaces for sodium, calcium and copper.

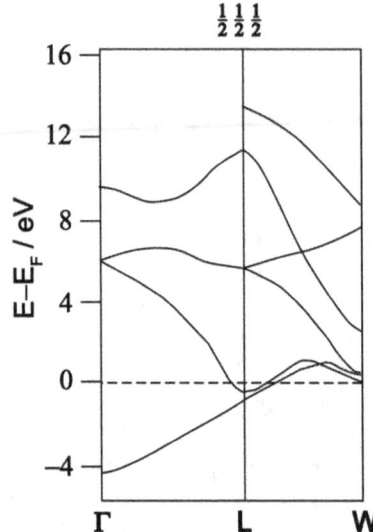

Figure 6.37. Computed band structure for calcium metal (adapted from [13]).

two electrons thus filling $2N_{states}$ available states. This would be expected therefore to fill the energy band right up to the band gap. Yet calcium is a metal! The answer comes from the fact that the Fermi surface is a three-dimensional object (figure 6.36). If we try to inscribe a sphere of the right size into the Brillouin zone shape, we discover that in certain directions (X and L) of the Brillouin zone the radius of the sphere is larger than the maximum values of **k** and this allows electrons to occupy states in the conduction band. This can be seen in the computed band structure (figure 6.37), which shows that the Fermi energy lies just above the band gap at the L point, with a very small band gap at W.

Diamond, silicon, germanium

These crystals all have the same basic crystal structural arrangement, yet diamond is an insulator while silicon and germanium are semiconductors. So, let's do the counting. The crystal structure is fcc with eight atoms in the unit cell. Therefore, in a primitive cell there must be only two atoms. Now, each atom contributes four valence electrons and so there are eight electrons per primitive cell. If we have N primitive unit cells in the crystal, then there are $2N_{states}$ in each band, and therefore four bands are filled at 0 K. Now in diamond there are extremely strong covalent bonds between the carbon atoms, and so the lattice potential is very deep at the atomic positions, thus making a very large band gap (approximately 5.5 eV). Diamond is therefore a good electrical insulator, even at room temperature, despite being an excellent thermal conductor. In silicon and germanium the bonding is less strong with smaller band gaps (1.11 and 0.67 eV, respectively). In these cases, input of thermal energy is sufficient to drive some of the electrons near the Fermi surface across the band gap to allow electrical conduction. It is therefore obvious that the conductivity will increase with temperature, and, despite Pauli's view, these materials are *intrinsic semiconductors*.

References

[1] Available https://chem.libretexts.org/Core/Physical_and_Theoretical_Chemistry/Statistical_Mechanics/Properties_and_Observables/Heat_Capacity_of_Solids

[2] http://en.wikipedia.org/wiki/Debye_model

[3] Glazer A M Monatomic and diatomic chains [Online]. Available http://www.amg122.com/programs/chains.html

[4] Brillouin L 1946 *Wave Propagation in Periodic Structures* (New York: McGraw-Hill)

[5] Wannier G W 1959 *Elements of Solid State Theory* (Cambridge: Cambridge University Press)

[6] Kittel C 2005 *Introduction to Solid State Physics* 8th edn (New York: Wiley)

[7] Kresse G, Furthmüller J and Hafner J 1995 *Ab initio* force constant approach to phonon dispersion relations of diamond and graphite *Eur. Lett.* **32** 729

[8] Bradley C J and Cracknell A P 1972 *The mathematical theory of symmetry in solids: representation theory for point groups and space groups* (Oxford: Oxford University Press)

[9] Cracknell A P 1974 *Group Theory in Solid-State Physics* (London: Taylor and Francis)

[10] Colthup N B, Daly L H and Wiberley S E 1990 *Introduction to Infrared and Raman Spectroscopy* (New York: Academic Press)

[11] Furrer A, Strässle T and Mesot J 2009 *Neutron Scattering in Condensed Matter Physics* (Singapore: World Scientific)

[12] Beauvois K and Klicpera M 2015 Inelastic neutron scattering Part 1: triple axis spectrometer *What do we measure in inelastic neutron scattering?* (Grenoble: Institut Laue-Langevin)

[13] Chatterjee S and Chakraborti D K 1971 Energy band structure of Ca, Sr and Ba *J. Phys. F Met. Phys.* **1** 638

IOP Concise Physics

A Journey into Reciprocal Space
A crystallographer's perspective
A M Glazer

Appendix
Wigner–Seitz constructions

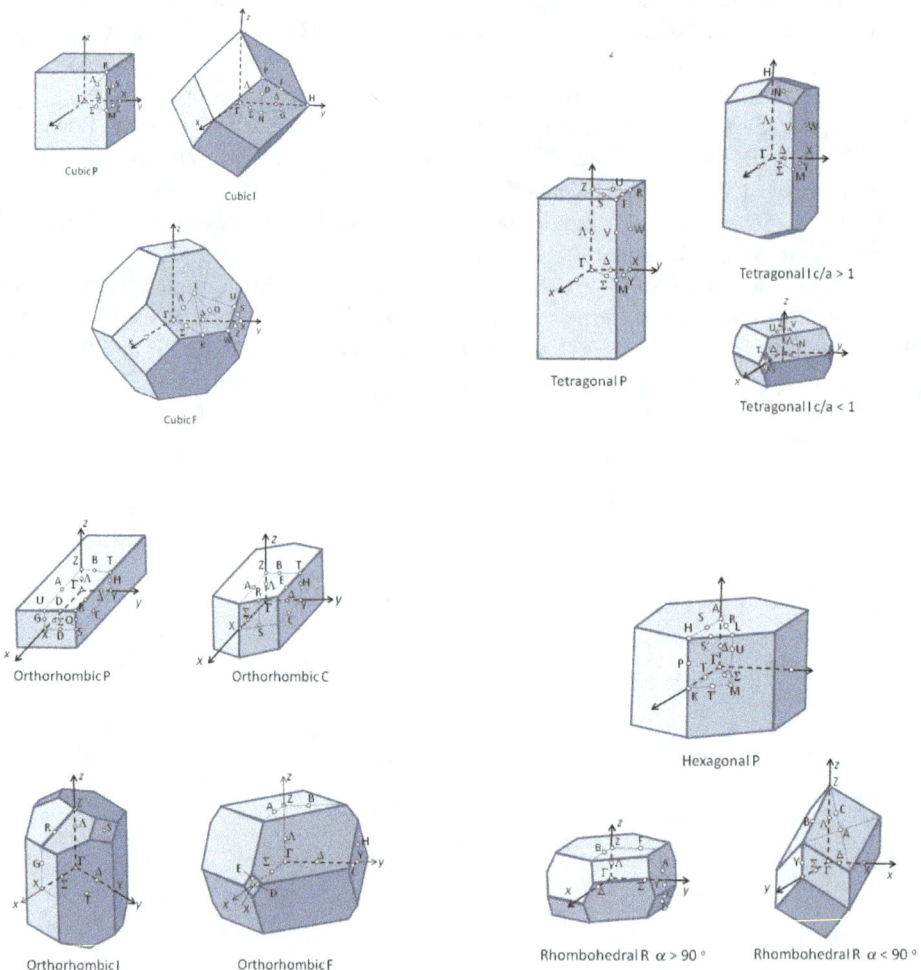

A Journey into Reciprocal Space

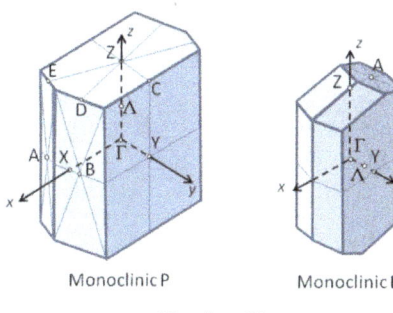

Monoclinic P   Monoclinic B

(Continued.)

www.ingramcontent.com/pod-product-compliance
Lightning Source LLC
Chambersburg PA
CBHW081428220526
45466CB00008B/2307

9 781681 746203